Collins

simple
stargazing

For Morten, Etienne and Dad

simple
stargazing

a friendly handbook for viewing the Universe

ANTON VAMPLEW

Collins

HarperCollins Publishers Ltd.
77-85 Fulham Palace Road
London
w6 8jb

The Collins website address is:
www.collins.co.uk

Collins is a registered trademark of
HarperCollins Publishers Ltd.

First published in 2005

10 09 08 07
10 9 8 7 6 5 4 3

A catalogue record for this book is available from the British Library.

ISBN-13: 978 0 00 720395 6
ISBN-10: 0 00 720395 0

Commissioned by Helen Brocklehurst
Edited by Caroline Taggart
Design and layout by Richard Marston
Proofread by Kate Parker
Index by Hilary Bird

Colour reproduction by Colourscan, Singapore
Printed and bound in China

Collins

Contents

Introduction

Prepare yourself for an adventure... that will take you deep into space and far back in time. This great journey begins the moment you cast your eyes up into the night sky. After a while you'll be looking beyond the stars, wondering about distant life or maybe thinking just how fantastically big this whole Universe thing is.

I vividly remember when I was six, gazing out of my bedroom window with a desire to learn the names of the bright stars and the patterns I knew existed in the form of constellations. Little did I know what I had started – a lifelong trip which never ceases to amaze me. We are now in the age of Hubble, Cassini, Galileo, Hipparchos and Messenger (etc., etc.) – spacecraft that open up the vistas of the Universe to realms that excite while causing us to constantly rearrange our jigsaw of... well, everything. And this is not going to slow down. Look out for specially trained astronauts on the Moon and Mars, and 'ordinary' astronauts (that's you and me) taking short trips into space.

Hopefully I can share some of my wonderment through these pages. None of this is rocket science (apart from the rocket science stuff). The name of the stargazing game is easy, short observing whenever you have a few spare moments while the stars are twinkling overhead.

'Oh, but I can't see the stars from where I live,' is always a good one. Read carefully, as this will be written only once: living in a town, city or anywhere with light-polluted skies need not deter anyone from stargazing. Although the sky glow washes out the fainter stars, the major constellations will still be visible. So you won't be hindered from learning the main star patterns. No excuse there, then!

Don't underestimate the power of 'doing' something. Simply by taking a few minutes each day over the course of a year you'll soon be amazing your friends as you point out Leo and say, 'Of course, Regulus is a B7-type star about 85 light-years away.' Or you might glance at the Square of Pegasus, remarking casually, 'Messier 15 over there was discovered by the wonderful Italian, Maraldi.' Or even dreamily waft your hand in the direction of Orion, and with a certain authority launch into, 'The dimensions of M42 are 66 by 60 arc minutes.' It won't take long to learn the night sky, and I hope this book will inspire you to make a start.

The night sky is out there. As this old map shows, it's all prepared and ready for you to explore.

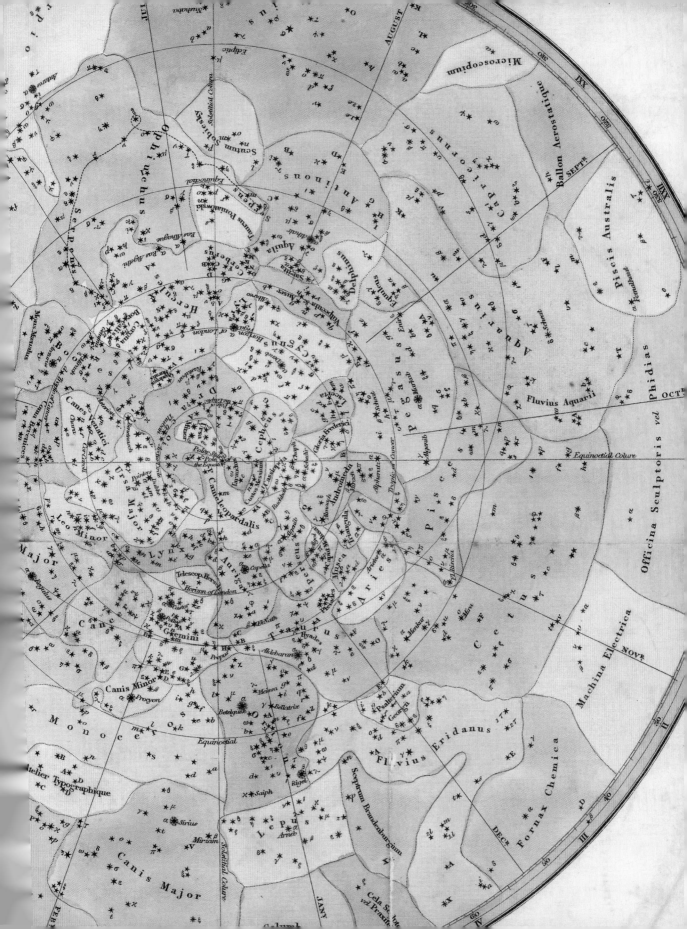

A fine sunset is worth a picture itself as well as the hint that it's going to be a fine, clear, starry-skied evening. Just the prompt you need to get you into the stargazing frame of mind.

Getting Started

A Brief History

In the distant past, astronomy and astrology were as one. Ancient rulers needed to know their fortune and, as the sky was where their gods lived, it was also where their destiny lay. Along with all the 'fixed' stars of the constellations were seven things that moved: the Sun, Moon and five planets – Mercury, Venus, Mars, Jupiter and Saturn (this was, of course, in the days when everyone believed that the Earth was the centre of the Universe and other sky objects moved round it).

It was an absolute belief that leaders who could understand how these objects moved could stay in control and defeat their enemies. One thing was clear: to these ancient watchers of the skies these seven objects followed a 'path' around the heavens – just like a car on a race track that takes the same route round again and again. It was the constellations situated along this 'path' that became our 12 famous signs of the zodiac.

Of course, in order to know where any object would be in the zodiac at any given time, a certain amount of calculation was required. This is when the science of astronomy was born. So, strangely, the necessity for fortune-telling encouraged the formation of science. By the way, *zodiac* means 'line of animals' (11 of the original 12 constellations are still animals) and is also linked to the word zoo.

So, why do the planets, Sun and Moon appear to move through the skies? Well, they each appear to move for different reasons. Of course the main movement you see is due to the Earth spinning – this gives us things like sunset, the Full Moon rising over frosty trees, time for your cornflakes for breakfast as the Sun rises, etc. The Moon, if it is up, additionally appears to move *extremely* slowly hour by hour in front of the stars because it is orbiting the Earth. The Sun changes its position against the stars day by day due to the fact that we are orbiting it. The planets move because they too are orbiting the Sun – plus each planet is moving at a different speed. No wonder it was all difficult to calculate, and indeed it's hardly surprising that some early astronomers ended up having their heads chopped off, when their erroneous adding up was followed by a total overreaction from their bad-tempered rulers.

Constellations

A word worth defining before we launch ourselves into space is *constellation*. It's based on a word from Latin meaning 'group of stars'. In total you'll find 88 of them filling the entire sky, but thankfully you don't need to know them all to enjoy the hours of darkness. Other starry terms that crop up throughout the book are written in **bold** and explained in the AstroGlossary on page 151.

The story of organising things up there in the darkness of the night began thousands of years ago with civilisations such as the Sumerians, Babylonians, Egyptians, Greeks and Romans (as well as many other cultures from around the world). They decided the starry skies could do with a bit of order and a tidy up. So they joined up many of the stars, just like a dot-to-dot picture, putting their myths and legends into the sky as they did so.

Don't think that there was any rhyme or reason for making a particular pattern. For example, Cepheus, King of Ethiopia, and his wife, Queen Cassiopeia, both have constellations named after them, and yet these look like a house and a set of stairs respectively. Imagination is the key here, I feel. As far as these early civilisations were concerned, the gods and goddesses needed a place to reside in the starry vault, so it was probably a case of first come, first served, and pot luck as to which stars were assigned to which group.

We get our earliest knowledge of the constellations from Aratos, the first Greek astronomical poet, in his work *Phaenomena* (which was probably based on an earlier 'lost' work by another Greek, Eudoxus). Then in AD 150 Ptolemy, a Greek working at the great library of Alexandria in Egypt, recorded them in a book known by its Arabic name, *Almagest*, which means 'the greatest'. Hundreds of years ago, other astronomers who wanted to be famous added extra groups (some more successfully than others) to give us our present fixed total of 88 constellations.

Constellation names are traditionally written in Latin. This is because Ptolemy's book was brought from the Middle East to Italy, where it was translated – and Latin, for centuries, was the language of scholars. So, for example, we know the Great Bear as Ursa Major.

Here are all the 88 constellations of the starry skies. Details of those with interesting things to see are given in Parts 2 and 3.

Latin Name	English Name	Abbreviation	Order of Size (1 is the largest)
Andromeda	Andromeda	And	9
Antlia	Pump	Ant	62
Apus	Bee	Aps	67
Aquarius	Water Bearer	Aqr	10
Aquila	Eagle	Aql	22
Ara	Altar	Ara	63
Aries	Ram	Ari	39
Auriga	Charioteer	Aur	21
Boötes	Herdsman	Boo	13
Caelum	Sculptor's Tool	Cae	81
Camelopardalis	Giraffe	Cam	18
Cancer	Crab	Cnc	31
Canes Venatici	Hunting Dogs	CVn	38
Canis Major	Great Dog	CMa	43
Canis Minor	Little Dog	CMi	71
Capricornus	Sea Goat	Cap	40
Carina	Keel	Car	34
Cassiopeia	Queen	Cas	25
Centaurus	Centaur	Cen	9
Cepheus	King	Cep	27
Cetus	Whale	Cet	4
Chameleon	Chameleon	Cha	79
Circinus	Drawing Compass	Cir	85
Columba	Dove	Col	54
Coma Berenices	Berenice's Hair	Com	42
Corona Australis	Southern Crown	CrA	80
Corona Borealis	Northern Crown	CrB	73
Corvus	Crow	CrV	70
Crater	Cup	Crt	53
Crux	Cross	Cru	88
Cygnus	Swan	Cyg	16
Delphinus	Dolphin	Del	69

Dorado	Goldfish	Dor	72	**Pictor**	Painter	Pic	59
Draco	Dragon	Dra	8	**Pisces**	Fish	Psc	14
Equuleus	Little Horse	Equ	87	**Piscis Austrinus**	Southern Fish	PsA	60
Eridanus	River	Eri	6	**Puppis**	Stern	Pup	20
Fornax	Furnace	For	41	**Pyxis**	Compass	Pyx	65
Gemini	Twins	Gem	30	**Reticulum**	Net	Ret	82
Grus	Crane	Gru	45	**Sagitta**	Arrow	Sge	86
Hercules	Hercules	Her	5	**Sagittarius**	Archer	Sgr	15
Horologium	Clock	Hor	58	**Scorpius**	Scorpion	Sco	33
Hydra	Water Snake	Hya	1	**Sculptor**	Sculptor	Scl	36
Hydrus	Little Snake	Hyi	61	**Scutum**	Shield	Sct	84
Indus	Indian	Ind	49	**Serpens**	Serpent	Ser	23
Lacerta	Lizard	Lac	68	**Sextans**	Sextant	Sex	47
Leo	Lion	Leo	12	**Taurus**	Bull	Tau	17
Leo Minor	Little Lion	Lmi	64	**Telescopium**	Telescope	Tel	57
Lepus	Hare	Lep	51	**Triangulum**	Triangle	Tri	78
Libra	Scales	Lib	29	**Triangulum Australe**	Southern Triangle	TrA	83
Lupus	Wolf	Lup	46	**Tucana**	Toucan	Tuc	48
Lynx	Lynx	Lyn	28	**Ursa Major**	Great Bear	UMa	3
Lyra	Harp	Lyr	52	**Ursa Minor**	Little Bear	UMi	56
Mensa	Table	Men	75	**Vela**	Sails	Vel	32
Microscopium	Microscope	Mic	66	**Virgo**	Maiden	Vir	2
Monoceros	Unicorn	Mon	35	**Volans**	Flying Fish	Vol	76
Musca	Fly	Mus	77	**Vulpecula**	Fox	Vul	55
Norma	Level	Nor	74				
Octans	Octant	Oct	50				
Ophiuchus	Serpent Bearer	Oph	11				
Orion	Hunter	Ori	26				
Pavo	Peacock	Pav	44				
Pegasus	Flying Horse	Peg	7				
Perseus	Perseus	Per	24				
Phoenix	Phoenix	Phe	37				

Adventures in Darkness

Right, you've opened the door and are standing in the garden/yard/field/outback/ savanna/rocky landscape/swamp, etc. gazing up at the night sky searching for something wonderful to appear. How many stars can you see on a clear night? Millions? Squillions? Zillions? In fact, away from light pollution, with a good low horizon, *the maximum number of stars you can see at any one time is 4,500(ish).* Count them if you don't believe me. Of course, if you live in a major city, then bright orange skies can easily reduce this number to less than 200, so the darker your location the better.

A few things to start with…

Step-by-step guide to stargazing

1

Before you go out, check where the Sun rises and sets from where you live. This will give you some idea of where to look when trying to find something in the night sky. Usefully, around 21 March and 23 September, the Sun rises exactly east and sets exactly west. However, in the northern hemisphere during the summer months the Sun appears (roughly and depending on the precise date) somewhere from the northeast and sets somewhere northwest, while in the winter it's a southeast rise and a sort-of southwest set. In the southern hemisphere, the summer Sun rises somewhere in the southeast and sets somewhere southwest, while winter sees a northeast rise and a northwest set.

2

In order to see the most stars you need to let your eyes become accustomed to the dark. This is called dark adaptation. Ten minutes is a good time to sit in the dark without the lights on. Ponder, cogitate and muse over the wondrous spectacle that you are about to marvel at. How many constellations will you find? This process of dark adaptation not only widens your pupils to let in more light, but also allows various chemical reactions to take place in your eyes and activate your light-recepting rod cells. Now you will be able to see all those faint stars.

Help your dark-adapted eyes by making sure any torch is covered in red plastic.

3

Whilst outside in the dark the only way to see where you're going, or to look at the great star charts in this book, is with a torch. However many you decide you need, each should be covered in red plastic or something similar. The resulting red light, you see, hardly affects your now dark-adapted eyes.

4

Grasp this book firmly and, if you are not one yourself, find a responsible adult and venture outside. Adults are very useful indeed for chatting to and for having someone who will marvel at your initial determination.

Where exactly to begin up there depends on where you live down here. For those in northerly climes, you need to go here (page 15) ... while southerly humans go here (page 18)...

Empty space – or is it?

Northerly humans start here

Just look at the page opposite: it's covered in what looks like a chaotic pattern of differently sized dots. Nothing could be further from the truth. Each dot is actually a star we can see in the night sky and, just like many things that look like chaos to start with, there's order within this mess.

Lurking within these dots you will find a very useful pattern that is probably the best place to start your stargazing quest in the northern hemisphere. This group is known affectionately as *the Plough* – well, it is in Britain. Moving around our planet, we find that the Plough is called Karlsvogna (Carl's Wagon) in Norway, the Big Dipper in the USA and the Saucepan in parts of France. This is definitely a good name for the shape, as you can see – a pan with a handle stretching out to the left. Anyone for space beans?

Now, the Plough is not actually a constellation itself, but part of a much larger group called Ursa Major, the Great Bear – we shall meet it very shortly.

The Plough is always visible from mid-northern latitudes if the skies are dark and the weather crisp and even. Also, all of its seven stars are quite bright, making it an easy group to find. In order to know which direction to look to find The Plough, you need to have some idea of north, south, east and west. As I said a moment ago, the Sun sets in the west(ish), so look to the right of that and up a bit (that's a technical term) and there's the Plough in the north(ish). Easy.

It's not long before patterns begin to emerge from the ether. Ether is an old term for the stuff that scientists used to believe filled space – it doesn't exist, but the idea's nice.

Round and round the Plough goes. If you are far enough into the northern hemisphere, this is where you'll find it at 8 p.m.(ish) at certain times through the year. The left of the diagram is the direction of northwest, whilst the right is northeast.

The Pointers of The Plough doing their 'pointing' thing.

Because the Earth is constantly turning, don't expect the Plough to stay in the same place for long. There's also our movement around the Sun to consider, which means that each night at the same time the Plough will be in a slightly different position. How exciting is that!? Generally you'll find the Plough higher in the sky during spring and summer evenings, and nearer the horizon in autumn and winter evenings.

As you may have noticed, there is a well-known star 'locked' in the centre of the image that the Plough rotates around. This is *Polaris*, also known as the *North Star*, or indeed the **Pole Star**. This last name means that it is the closest star to the North (Celestial) Pole, but because of the way the Earth spins on its axis, this is a temporary title and has been held over millennia by a number of the stars featured in this book.

You can always find Polaris by using the two right-hand stars of the Plough, which are called the Pointers. No need for Sherlock Holmes here – these two stars, Dubhe and Merak, simply 'point' up out of the 'saucepan' to Polaris. Elementary. And this is just one reason

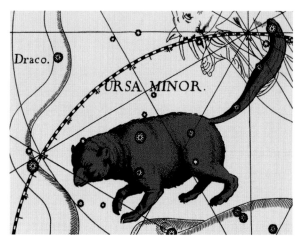

Polaris is the leading (main) star of Ursa Minor, the Little Bear.

why the Plough is so useful. Through this book you'll find plenty of ways that it can be used as a 'signpost' to many other stars and constellations.

Now, to break a myth: *the North Star is not the brightest star in the night sky*. It seems that for some unknown reason someone, somewhere, sometime told us that not only was it the brightest but it was also the first star you could see when it got dark. This is not true: Polaris happens to be only the 50th brightest star in our skies. Its fame is due to its position: almost directly above the North Pole. As the Earth spins we see the effect as the sky spinning, and in the northern hemisphere it's Polaris that everything goes around. Being almost stationary in the sky means that if you're looking at it you are looking north. And if you know where north is, you also know where east, south and west are. This is why Polaris was great in the olden days when mariners would 'sail by the stars'.

There's another group that can be found with supernoval ease by carrying on the line from the Pointers through Polaris to a 'W' shape that is Cassiopeia, the Queen. If your house/flat/hut/cave/tent/treehouse, etc. is in a position where the Plough never sets, then neither does Cassiopeia – they'll both be up, somewhere. Because they are on opposite sides of Polaris, when the Plough is high, Cassiopeia is low and vice versa.

Following imaginary lines made by stars can lead you anywhere in the Universe.

Cassiopeia, the Queen, sits and ponders: 'Hmmm, I know I've forgotten something?'

Southerly humans read this

Travelling to the southern part of the world, where the Plough may only be visible for half an hour in mid-April, or indeed may have totally vanished below the horizon, we need something else that can help us on our stargazing travels. Indeed, as for seeing the Plough (even for the briefest of periods), places near 23°S, like Alice Springs, Australia, São Paolo, Brazil, or Gaborone, Botswana, are really your most southerly locations.

A comparison in size between the Plough, a part of Ursa Major, and Crux, the Southern Cross.

What we're looking for in the southern skies is a small constellation, the smallest in fact, known as Crux, the Southern Cross.

Of course, as with everything else, the years have performed transformations and rearrangements of the part of the sky where the Cross that we know today appears. For example, in Ptolemy's day – the second century AD – the stars of Crux were part of the next-door constellation Centaurus, the Centaur. It was only in the late 16th century that the Cross began to take on its own personality as modern astronomers placed it in their star atlases.

Another change of names involves something you can see – or not see! – within the borders of Crux: a cloud of blackening dust

and gas which obscures the Milky Way stars behind (we'll be hearing a lot more about the Milky Way later in the book). Known today as the *Coal Sack*, in history this cloud has also been the *Soot Bag* and the *Black Magellanic Cloud* – which is a mysterious name I really like. It was once darkly described as 'the inky spot – an opening into the awful solitude of unoccupied space'.

Crux, and some friends that we shall meet very shortly, are the southern equivalent of the Plough and Pole Star combined, because they too can be used to find your way about in the dark. By following various imaginary lines you can fairly easily discover the South Pole of the sky – the point about which the stars seem to revolve.

Unfortunately when you get to this point, darkness prevails, for there is no star equivalent of Polaris awaiting your arrival – there is no South Star, or Polaris Australis, as you could have called it. Astronomers with big telescopes who do not wash much will harp on about the star σ Octantis, which almost marks the southernmost point. However, it is extremely faint, difficult to find and therefore almost useless. So the Southern Cross and a couple of dazzlers next door do the admirable job of locating this starry (or *celestial*) pole of the south.

As the Earth spins and travels around the Sun, you'll find Crux in different parts of the sky depending on the time and date. Its highest appearances occur during the autumn and

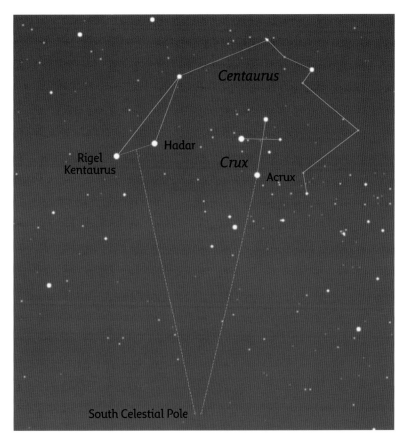

Round and round the Southern Cross goes. Using the Crux 'Pointers' and some useful jiggery-pokery with Rigel Kentaurus and Hadar, you easily can find the South Pole of the sky!

winter evenings; in spring and summer evenings it's nearer the horizon.

You can just glimpse a few stars of Crux from the Canary Islands off the north coast of Africa, but you'll need to go below about 23°N – to Aswan, Egypt, Hong Kong or Dacca, Bangladesh, for example – to see it in all its glory. If your latitude is further south than about 34°S, like Sydney, Australia, Montevideo, Uruguay, or Cape Town, South Africa, then technically the Southern Cross never sets – although it still just skims the horizon until you go even further south, which you'll have to do by boat as you run out of land!

Anyway, that's the 'where' bit. Next, what about how big things are?

The appearance of the Plough we know and love really depends on where we live. In the northern hemisphere it may be visible whenever it's dark. However, the further south you go, the less you see of it. Around 23°S it appears low over the northern horizon only during evenings in April – and even then it's upside down!

Travels into the Darkness

How big is space itself? The large distances on Earth still amaze me, let alone trying to imagine the great gaps between the planets. It's worth just a thought or two – see how much distance you can imagine. Take my house, for example: I have to walk about 1 km to get from there to the cake shop. That's a nice, easy stroll that takes me 10 minutes; I can picture that. Now the Moon, our nearest neighbour in space, is 384,000 times further than the cake shop. That is, of course, 384,000 km. Walking there would take me nearly nine years – and yet the Moon is only next door as far as space is concerned.

I'm already having a slight problem trying to imagine this relatively tiny Earth-to-Moon distance, so what chance do I have with larger gaps? For example, the distance from my house to the Sun is a massive 150 million km – that's already getting pretty big and we haven't left our solar system. The nearest star after the Sun, called Proxima Centauri, is about 40 trillion km from my front door and, by moving deeper into space, we can find the Andromeda galaxy, a *close* star system that is 26 *quintillion* km away!

26,000,000,000,000,000,000

And still these biggish numbers are just peanuts compared to the size of distances in the Universe – there really is a lot of space out there.

What does a quintillion mean to you? I have to say it doesn't mean much to me. So, if I'm having trouble with the distance to the Moon, what hope do I have with 26 of these quintillion thingies?

Help is at hand, though, as astronomers have a different way of measuring very large distances in space, and it's called the **light-year**. A light-year is simply the distance that light, zipping along at nearly 300,000 km

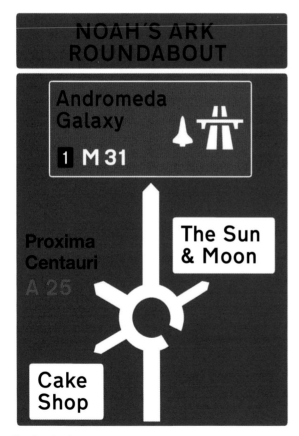

Your imagination can take you anywhere on the space super-highway. Then again, maybe this will become reality.

16 April 2002 at 20.55. The Moon and Saturn at the top, with the bright star Aldebaran at the bottom. All these objects look as if they're the same distance away from us, but Saturn is really 3792 times further away than the Moon, while Aldebaran is 911.5 million times more distant.

per second, travels in one year. Now, instead of our nearest star being 40 trillion km away, it becomes a more manageable 4.27 light-years.

Even so, the Universe as a whole space-time thingamajig is still a whopping 13.7 billion light-years across – something you should only try to convert into kilometres if you've got a *very* big piece of paper.

UNIVERSE CLOSED
13.7 BILLION LIGHT-YEARS AHEAD
(APPROX.)
WORMHOLE TRAFFIC ONLY

This is as far as we can go (at the moment!).

We can use the speed of light to measure times other than a year. Here's a table full of bits and bobs to give you some idea of the vastness of space:

Thing (planet times given are when they are at their closest to Earth)	One-way Light-travel time from or to Earth
Moon	1.25 seconds
Venus	2.3 minutes
Mars	4.35 minutes
Sun	8.3 minutes
Pluto	5.3 hours
Voyager II (furthest spacecraft as of 2004)	1 day
Proxima Centauri (our nearest star after the Sun)	4.27 years
Deneb (main star of Cygnus, the Swan)	~2,100 years*

* '~' means approximately, and is also used in the constellation of Cygnus.

Anyway, let's now amaze ourselves with just how big the 'space' you can see up there is...

How Big is the Darkness?

Depending on the hemisphere you are in, either the Plough or the Southern Cross is easy to spot if you know in which direction to look and how big they are. This idea of size is useful to understand, so I'm going to take a moment to show you how to measure things in the sky.

Let's start with the Moon. Most people would say that it is a lot bigger than it really looks. You may be surprised when you realise that the end of your little finger held at arm's length easily covers the Moon – with room to spare. Have a go next time the Moon is out.

Of course, you can cover different amounts of the sky using more of your hand, arm or even your feet if you're fit enough. For now it is useful to know that the Plough, as viewed from the Earth, is slightly longer than your outstretched hand held at arm's length. However, there are loads of tiny things to see, so it is time to get a little more scientific.

You probably know that if we want to divide any circle into smaller units, we use degrees – or, more accurately, angular degrees – and that 360 of them make up a full circle. If you imagine the circle as a clock, the minute hand moves through 360 degrees when it goes all the way round, which takes one hour.

A single degree is a very small measurement, equal to the barely visible movement that the minute hand on a clock makes in 10 seconds. But in space many objects are extremely small, so we need incredibly small units to measure with. The space boffins have therefore divided the degree into 60 smaller segments, and each one of those into a further 60 even smaller segments.

Unfortunately the names of these smaller segments sometimes lead to confusion, because the 60 smaller segments of an angular degree are called angular minutes (or arc minutes) and the 60 smaller segments of the angular minute are angular seconds (or arc seconds).

These units are represented by the following symbols:

° degrees   ' minutes   " seconds

The Plough appears only slightly longer than your outstretched hand – though it depends on how big your hands are, of course.

You can avoid confusion simply by remembering that if you see the word *angular* or *arc* anywhere, the measurements are to do with angles, not time.

With all this information about measuring, let's take a look at the sizes of some space things in degrees, arc minutes and arc seconds:

Thing	Approximate angular size
Distance from the Pointers in the Plough to Polaris	28°
Length of the Plough	24°
Your outstretched hand at arm's length (roughly)	22°
Distance apart of the Crux Pointers	6°
Distance apart of the Pointers in the Plough	5°
Your forefinger at arm's length	1°
Your little finger at arm's length	½°
The Sun	½°
The Moon	½°
Distance of Ganymede from Jupiter (that's the brightest of the planet's main Moons)	6'
Resolution of the unaided eye (this means your eye can see two objects that are very close together as two objects rather than one.)	3' 25"
Maximum size of Venus	1'
Biggest crater on the Moon	1'
Your eye can see single objects as small as... (about)	1'

What's interesting from this table is that technically we can see at least one of the moons of Jupiter as well as the crescent-shape phase of Venus simply by gazing skyward with our unaided eyes (the 20–20 variety is required). However, in practice the usually super-brightness of Jupiter drowns out the fainter light from its moons, while the dazzling appearance of Venus does the same for its crescent.

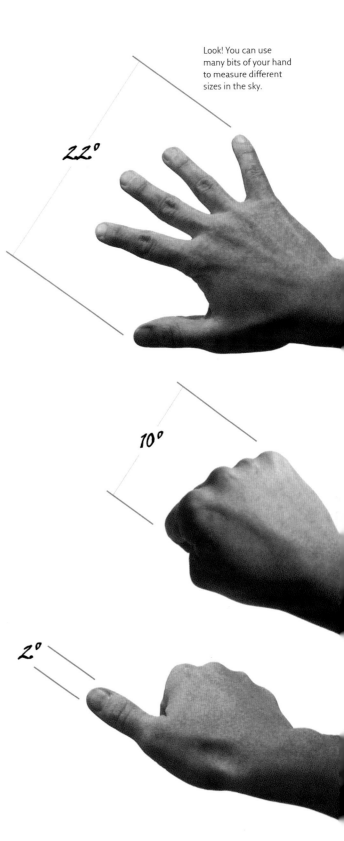

Look! You can use many bits of your hand to measure different sizes in the sky.

22°

10°

2°

How to Use
the Star Charts

What you can see depends on the time of year – the stars change from season to season due to our motion around the Sun. Therefore the constellations that we'll be discussing a bit later on have been divided into the skies of spring, summer, autumn and winter. Some constellations are visible all year, but they are *best* viewed at certain times of the year.

As we've seen with the Plough and Crux, several distinctively shaped and therefore easily recognisable constellations – or parts or combinations of constellations – make useful 'signposts' around the night sky. These can be used to find all manner of starry splendours, so I will point them out as we go.

The sections have also been divided into those for northerly people and those for southerly, because exactly what you'll see depends on whereabouts on our planet you live. If you're well into the northern hemisphere (the bit above the equator on the map) then go for the northern charts, while if you're in the southern... well, you get the idea. As most of the land – and the majority of the world's population – is in the northern hemisphere, the charts appear as if you are looking south. The closer to the equator you are in the northern hemisphere, the more of the southern chart you can use. For those lucky souls who live on the equator, the whole sky is visible and you can use both charts at once.

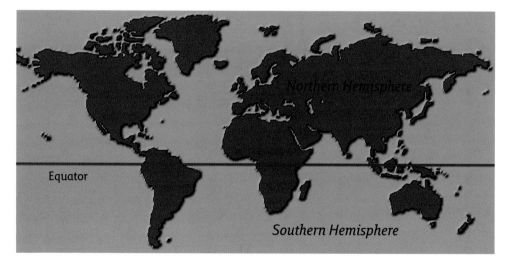

Equator

Northern Hemisphere

Southern Hemisphere

Here's a rough map of the planet that I knocked together using a plastic washing-up-liquid bottle, some sticky-backed plastic and an old coathanger. The equator divides the world into a northern and a southern hemisphere, which is your cue to choosing which of the seasonal charts to look at. The nearer to the equator you travel, the more of the opposite hemisphere's chart you can use. For example, if you live somewhere in Norway, you'll be using just the northern chart. If one summer you decide to go on holiday to Italy, you can use half of the southern chart with all its fascinating new starry sights.

Each constellation of interest is introduced with the following useful information:

LATIN NAME
Ursa Major
ENGLISH NAME
The Great Bear
ABBREVIATION
UMa
LATIN POSSESSIVE
Ursae Majoris

α STAR
Dubhe
MAGNITUDE
1.79
STAR COLOUR
Orange

First, you'll find the constellation names in the original *Latin*, and then the *English* equivalent. The following three-lettered *abbreviation* is an internationally recognised way of identifying a constellation without having to use its full name.

The *Latin possessive* means 'of the constellation' or 'belonging to'. This is used for sounding important and as if you know what you're talking about – something like: 'Oh, Castor. Of course you mean *alpha Geminorum*.'

The alpha (α) star is the main star of the named constellation, though it is not always the brightest and not all constellations have alpha stars with names (or even alpha stars at all). In 1603 the German astronomer Johann Bayer worked through each constellation, *usually* assigning alpha (α) to the brightest star, beta (β) to the next brightest, then gamma (γ), delta (δ) and so on. As a result, stars are often labelled on charts with a letter of the Greek alphabet, known as the Bayer Letter. The full Greek alphabet is given on the right.

Proper names, such as Dubhe, are only generally given to the brighter stars. Many of the names are Arabic, but there are splatterings of Greek and Roman ones too. Check out the constellation Libra (see page 99) for the best star names in the Universe.

Star magnitude (mag.), or 'visual magnitude', lets you know how bright a star appears in the sky. The Greek astronomer Hipparchus, who lived around the mid-second century BC, introduced the system of classifying starry magnitudes that you can see with your eye from 1, the brightest, down to 6, the faintest. We'll be talking more about magnitude in just a moment or two, and as you will see, it's now a little more complicated.

As well as the *colour* of a star indicating the obvious, it can also tell you how hot its surface is. You may be surprised to learn that the coolest stars are red, their surfaces being about 3,000°C. Warmer stars are yellow, the hotter ones white and the hottest are blue, with temperatures up to 40,000°C. Not only that, but they can change colour! This most notably happens when stars have used up all their fuel and internal forces take over, causing all manner of starry events (like red giants, supernovae and black holes, for example).

Greek alphabet

α	Alpha	η	Eta	ν	Nu	τ	Tau
β	Beta	θ	Theta	ξ	Xi	υ	Upsilon
γ	Gamma	ι	Iota	o	Omicron	φ	Phi
δ	Delta	κ	Kappa	π	Pi	χ	Chi
ε	Epsilon	λ	Lambda	ρ	Rho	ψ	Psi
ζ	Zeta	μ	Mu	σ	Sigma	ω	Omega

Bright or Dim?

As far as stars are concerned, the further away you put one the fainter it appears, just as a candle on a table next to you appears brighter than one on a nearby hill. Both candles are of the same brightness; it's just the distance you have to account for.

Now, what about a candle next to you and a great bonfire on that hill? These may *look* equally bright – in other words, they have the same *visual brightness*. However, because you are clever, you know that as you walked up the hill, the fire would appear brighter and brighter. So when we talk about visual brightness we mean how bright things are *from our viewpoint*, regardless of how far away they are.

It's the same with space, except that not only are the distances big, but the true brightness of stars and galaxies is incredible. The technical term for the actual brightness of something is **absolute magnitude** – you can say the candle on the hill and the one on the desk both have an absolute magnitude of 1 candle-power, but the visual brightness is greater for the nearby candle than the one on the hill.

On a more scientific level, we have measured the apparent brightness of objects in space - as they appear to us referred to by astronomers in the know - as **visual** or **apparent magnitude**, very accurately, and some of the closest are listed here:

Space object	Visual magnitude
	Brighter
The Sun	-26.7
The Moon	-12.6
Venus	-4.7
Mars	-2.9
Jupiter	-2.9
Mercury	-1.9
Sirius (brightest night-time star)	-1.4
Saturn	-0.3
Asteroid Vesta	5.3
Ganymede (Moon of Jupiter)	4.6
Uranus	5.5
Faintest unaided-eye objects	6.0
Neptune	7.7
Pluto	13.8
	Fainter

As shown on the chart, the faintest objects you can see on a super-clear night away from light-polluted skies are magnitude 6 and above, and as we count down to zero the objects get brighter. Then we head off into minus numbers for the brightest objects.

Interesting magnitude fact: it is said that humans can differentiate between just 0.1 of a magnitude. Have a go.

Here is a brightness guide to some of the stars around Orion, the Hunter. For an indication of how clear your skies are, see what magnitude you can get down to - you might be pleasantly surprised, or not, as the case may be.

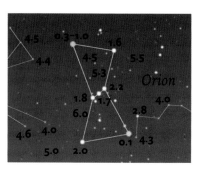

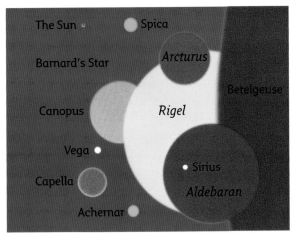

Here are some stars you will meet on your journey through this book. It's clear that stars really do come in a vast variety of sizes and colours. Look at tiny Barnard's Star, then compare it with mega-super-big Betelgeuse. In fact if we could put Betelgeuse where the Sun is, its 'surface' would be out near Jupiter's orbit! What sort of craziness is that!? What a star is like depends on how much gas made it in the first place and where it is in its life cycle. The sky is awash with these colours, but only the brightest stars have any hint of not being white because they are bright enough to trigger the colour receptors in our eyes.

Stars vary in distance from Earth as well as in brightness. Here's a guide to the 11 closest stars (I've included the Sun) and the 10 brightest stellar marvels. Visual magnitude in the table lets you know how bright the object looks in the sky and was chatted about fully earlier – I'm sure you were paying attention.

Nearest stars	Distance (light-years)	Visual magnitude	Constellation
The Sun	very close	-26.72	
Proxima Centauri	4.27	11.05	Centaurus
Rigel Kentaurus A	4.35	0.00	Centaurus
Rigel Kentaurus B	4.35	1.36	Centaurus
Barnard's Star	6.0	9.54	Ophiuchus
Wolf 359	7.8	13.45	Leo
Lalande 21185	8.3	7.49	Ursa Major
Sirius A	8.6	-1.46	Canis Major
Sirius B	8.6	8.44	Canis Major
UV Ceti A	8.7	12.56	Cetus
UV Ceti B	8.7	12.52	Cetus

Brightest stars	Distance (light-years)	Visual magnitude	Constellation
Sirius	8.6	-1.46	Canis Major
Canopus	313	-0.72	Carina
Arcturus	37	-0.04	Boötes
Rigel Kentaurus A	4.35	0.00	Centaurus
Vega	25	0.03	Lyra
Capella	42	0.08	Auriga
Rigel	773	0.12	Orion
Procyon	11	0.38	Canis Minor
Achernar	144	0.46	Eridanus
Betelgeuse	427	var. 0.30-1.00	Orion

Stars

Stars are the glittering jewels of the night sky. Big or small they hang about in space shining away until they stop shining and start behaving badly, like black holes. They're bad because gravity has taken over to the extreme: imagine a big super-powerful vacuum cleaner. You don't want to get too close to them, otherwise your atoms would get squashed and splattered all over the spaceship window - nasty! Then there are those stars that are beginning to become unhappy - like variable stars - whose stellar furnaces are unstable. With these, the signs of age are showing with size and brightness fluctuations that only look nice from a distance. You are well advised never to go near any old stars. This is all in contrast to our sun.

Our Sun, so far as we know, is on its own, which is quite rare as plenty of other stars like a bit of company. These are double stars. You may not notice this with your eye, but out there live families of stars. How can you tell what's what? With a bit of patient stargazing you can find some interesting stellar marvels.

Double stars

What is a double star? Simply, what may look like a single star can really be two or more stars sitting very close together in the night sky. They come in two varieties:

Optical doubles are not related in any way. These stars simply look close because of our viewpoint.

Some stars that look very close together are in reality not. They appear close because they happen to be in roughly the same direction as seen from the Earth. Take Dominic and Janet, for example: at the top is the view from your back window, but away from the Earth we can see they really are a long way apart.

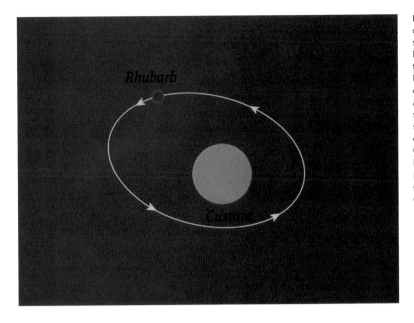

Here's one example of a binary star – two stars locked together by gravity. Orbits may take anything from hundreds of years down to a few hours. Also you may find several stars in the system – for example, alpha Geminorum (Castor) has six stars taking part in a gravitational orbital dance around each other!

Binary pairs really are at the same distance away from us, with the stars orbiting each other under the pull of gravity.

All the double stars throughout this book are shown to the same scale, so you'll know easily how one compares in size to another.

Top tips for double-star viewing

There are quite a few double stars lurking out there in the night sky, but some may be more of a challenge if your skies are light-polluted.

Your eye can see double stars as close as about 3' 25" apart. The more equal the brightness of the stars, the closer together they can be seen.

EXAMPLE

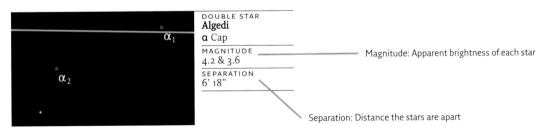

DOUBLE STAR
Algedi
α Cap

MAGNITUDE
4.2 & 3.6 — Magnitude: Apparent brightness of each star

SEPARATION
6' 18" — Separation: Distance the stars are apart

Variable stars

What is a variable star? It's one whose brightness changes over time. This can be for a variety of reasons: for instance, unstable old stars pulsating in and out, growing bigger and smaller, or close binary stars whose orbits mean that each star moves in front of the other as they go round. Some stars even swap gas, which causes big explosions called *novae*.

Variable stars are worth having the odd look at, if only to observe another ever-changing aspect of the starry skies.

The types of variable stars are named according to the first star identified as a new class. This list shows you a few examples of variable stars, their names and what they do – sometimes with explosive consequences.

Type of variable (some exciting names here!)	What that means, what it's like, what it does and why it does it
Mira	**Long period** An old red star that changes in brightness irregularly over periods that may range from hundreds of days to many years due to the unstable internal workings causing the star to pulsate in and out. The name derives from the first variable star to be identified (see page 79).
Cepheid	**Regular period** Another type where it's the insides that are causing the changes in the star. But Cepheids are a most regular type of variable – they pulsate in a very predictable manner. If you made a graph of time versus brightness, you would get a rise and fall (like a wave) over an exact period – some stars take a day, while others take a month or more to complete their cycle.
RR Lyr	**Short regular period** Yet another type of star with internal problems. Here the brightness can change over just a few hours, while the longest variance within this stellar group is only about one day.
γ Cas	**Irregular period** This type of eruptive variable fades and grows brighter because of violent goings-on in its atmosphere.
R CrB	**Unpredictable anti-nova** This is an older sun-like star whose atmosphere occasionally gets clogged with soot. The star fades slowly while this happens until the soot is 'coughed' out into space and the brightness increases to its normal level again.

Algol	**Eclipsing binary**

Eclipsing binary

Picture the scene: two stars orbiting each other. In fact, imagine the image from the double-star info (page 29) where Rhubarb and Custard actually appear to cross in front of each other. As we observe these 'starry eclipses' there is a change in brightness. When the stars are apart we see the light from the whole system, but when they are together there is a dip in brightness as part or all of the further star is 'eclipsed'. The periods of Algol-type variables are regular, being simply the time it takes for the stars to orbit each other.

T CrB	**Recurrent nova**

Recurrent nova

Again, two orbiting stars, but this time one is big and one is small. The smaller one pulls 'star stuff' off the big one until there's enough to explode – when we see a great brightening in the sky. These types are very irregular in period.

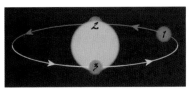

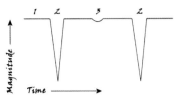

Here's how a star system, just like Algol, can change its brightness. The orbits of the stars around each other have been simplified to make things nice and easy. The graph shows how the brightness of the system changes as the stars go round each other. First imagine being so far away that these stars appear as one in the sky. When the small star is in position **1** then we see the combined light from both stars and our graph is at maximum (we're seeing the full brightness of the system). However, in position **2** the small is now semi-eclipsed: the graph plunges as we see a dip in the magnitude of the system. As the small star emerges the brightness returns to normal once more.

A while later (hours, days, weeks or years!) the small star moves in front of its larger companion, as shown in position **3**. This time we don't see such a great decrease in magnitude as it is only one small part of the larger star's light that is being blocked. Of course this is an example of only one sort of alignment. In some systems one star can completely hide the other, whilst the sizes of the stars themselves can be vastly different – you may get two (or more!) the same size orbiting in a spinning dance of stellar gravitational craziness. Over time we can see and understand the rhythmic nature of the variable starry heavens by studying these light-curves – some of which are extremely complicated.

Remember: these stars are variable and as such their atmospheres and internal structure can rise and fall, but they will not get back all the material lost. A star is at risk if it does not keep up the thermal pressure required to keep it gravitationally stable. Stellar evolution is protected by the laws of physics as we know them. Written quotations available on request.

EXAMPLE

VARIABLE STAR
δ Cep

TYPE
Algol eclipsing binary

MAGNITUDE RANGE
3.5 & 4.4

PERIOD
5.3663 days

Type: What kind of variable star is it?

Magnitude range: Brightness change of the star

Period: Time over which brightness changes

Top tips for variable-star viewing

When viewing variable stars, remember that the faintest star you can see with the naked eye in perfect conditions – i.e. not in a town – is magnitude 6. So when one seems to have disappeared, it's probably because it is temporarily not bright enough to be seen by the unaided eye.

It may also be interesting to note when such a star becomes too faint to see from your location – this would give you some idea of how clear your skies are.

Starry Objects

The night sky is full of deep-sky objects. These exist far outside our solar system and come in a range of styles and colours. On the charts the following symbols are used as positioning markers for some cosmic unaided-eye marvels:

Symbol   **What that means** (in a nutshell)

 Galactic cluster: Lots of younger stars in a group

Globular cluster: Lots and lots of stars in a ball-shaped group

Nebula: A cloud of dust and gas illuminated by stars

Galaxy: Lots and lots and lots and lots of stars

Let's expand these terms so you know exactly what's going on out there...

Galactic clusters

The first variety of starry cluster we meet is the *galactic*. These families of anywhere from a few dozen to a few thousand stars are born in the dusty spiral arms of our galaxy. They are basically new stars travelling together, but gentle tidal forces will eventually cause the stars to move apart until each stand alone in the night sky. Fine examples include the Pleiades in Taurus and the Jewel Box in Crux.

The Trapezium galactic star cluster is in the heart of the Orion nebula, shown in its full glory on the next page. (A Hubble image courtesy AURA/ STScI / NASA)

Gobular clusters

The second type of star cluster is the *globular*. These are much bigger than the galactic sort, consisting of hundreds of thousands to millions of generally reddish older stars. Whereas galactic clusters are found within our galaxy (hence the name), the globulars form a halo around it. Unaided-eye examples include the Great Globular in Hercules and Omega Centauri in Centaurus.

The M80 globular cluster in Scorpius. (A Hubble image courtesy AURA/ STScI/NASA)

The Orion nebula (M42), an emission nebula from the next section, has trespassed into the galactic clusters arena. Reason: well, space is just a mish-mash of things that very often link up with other objects.

Basically, one of the stars from the Trapezium cluster (see left) is responsible for lighting up most of this wonderful cloud of dust and gas. (Image courtesy John Punnett/Greg Smye-Rumsby)

Nebulae

Over the centuries, observers have noticed several small, faint, almost cloud-like patches between the stars, simply by gazing up into the night sky. These were given the name *nebulae*, which is Latin for 'clouds', due to their wispy appearance.

Because nobody knew exactly what was going on in the nebulae they were unaware of the true nature of these clouds. When telescopes became large enough, some of the nebulae were revealed to be galaxies, as was the case with the Andromeda nebula, now known as the Andromeda galaxy. Others were confirmed as real nebulae - areas of dust and gas. They are classified as follows:

Emission nebulae These are the brightest type of nebulae, which shine because of hot stars embedded in them. The radiation from these stars excites the surrounding gases, causing them to glow. Emission nebulae can be enormous; in fact there's enough gas and dust in them to make stars and planets - they are stellar nurseries. One of the easiest to see with the unaided eye is M42, the Orion nebula.

Reflection nebulae As the name suggests, these 'clouds' are visible because they simply reflect the light of nearby stars. The stars cannot cause the gas to glow because they are cooler and not so energetic, with the result that reflection nebulae are much fainter. The Pleiades, in Taurus, are surrounded by a faint nebula, but it can only be seen with more powerful telescopes.

Dark nebulae Gas and dust with no nearby stars do not shine, and we can only see them because they block out the light of anything behind. Included in this category are the famous Horsehead nebula in Orion - you'll need a telescope - and the much larger Coal Sack in Crux of the southern hemisphere 'easy to see with the naked eye'.

Planetary nebulae Some stars puff off their outer layers at the end of their lives, leaving a small, hot, energetic star. The layers expand outwards, shining with radiation from the central star, in a similar fashion to emission nebulae. Through a telescope this 'shell' looks somewhat like a planet, hence the name. The Ring nebula in Lyra is the classic example.

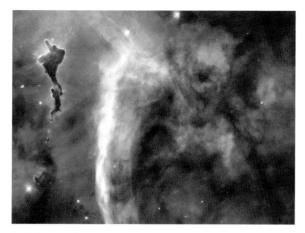

Emission nebulae
The dark Keyhole
nebula (on the left)
sits in the amazing
emission nebula Eta
Carina. (A Hubble
image courtesy
AURA/STScI/NASA)

Dark nebulae
Part of the dark
Horsehead nebula in
Orion. (A Hubble
image courtesy
AURA/STScI/NASA)

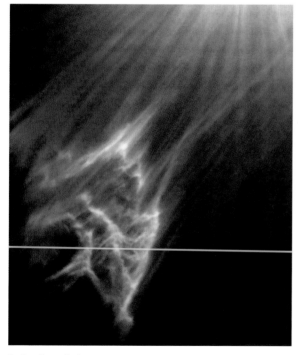

Reflection nebulae
It's behind you! The
ghostly reflection
nebula IC 349 in the
Pleiades. (A Hubble
image courtesy
AURA/STScI/NASA)

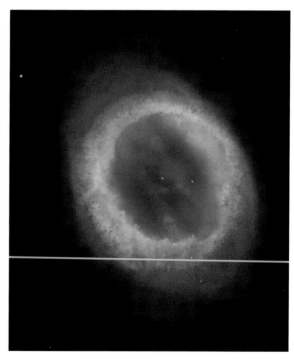

Planetary nebulae
The stunning Ring
nebula, M57, in Lyra.
(A Hubble image
courtesy AURA/
STScI/NASA)

Galaxies

There is a wild assortment of galaxies beyond our own: some round, some squashed, some fat, some thin, some big, some small, some beautifully formed and some bizarrely shaped. At the end of the day, they come down to just a few basic types:

Spiral galaxies These have a central bulge of older stars from which begin the great spiral arms of youngish stars, dust and gas. We live in a spiral galaxy (known dramatically as the Galaxy) and another is the furthest object we can see with the unaided eye, M31, the Great Andromeda galaxy, sitting over 2.8 billion light-years away.

Barred spiral galaxies These are like spirals but have a (guess what?) bar feature stretching out on each side of the central bulge. Spiral arms then start from each end of this bar. Some theories, and there are plenty more of them, suggest that the rotation of the galaxy itself may cause the bar, as the stars just line up for a while before the rotation disperses the feature.

Elliptical galaxies These have no spiral structure and not a lot of dust and gas. They can be small or enormous, the largest galaxies we can see, in fact. Because of their size some astronomers believe that elliptical galaxies may be formed by collisions of spiral galaxies; others say spirals evolve from ellipticals – this leads to unpleasant scenes in lots of observatories.

Irregular galaxies Generally small and faint, these have no structure whatsoever. Each is a hotch-potch of stars and gas that has been through a mid-life crisis. Collisions between galaxies may have been involved, but whatever the case... the truth is out there.

Spiral galaxy
NGC 4414 in Canes
Venatici (A Hubble
image courtesy
AURA/STScI/NASA)

**Barred spiral
galaxy** NGC 1300 in
Eridanus (A hubble
image courtesy
AURA/STScI/NASA)

Elliptical galaxy
M87, an elliptical
galaxy in Virgo
(Image courtesy
Anglo-Australian
Telescope)

Irregular galaxy
M82, in Ursa Major
(Image courtesy
AURA/NOAO/NSF)

Starry object catalogues

On the star charts and within the text you'll find lots of deep-sky objects with names like M4 and NGC 664. These come from a variety of catalogues that have been crafted for your viewing pleasure over many centuries. Here's a quick guide to those used:

M is the *Messier Catalogue*. Charles Joseph Messier (1730–1817) was fed up of confusing 'fuzzy' objects with the comets he was trying to discover. So in 1781 he put together this collection of mostly nebulae, galaxies and star clusters. Many of them are visible with the naked eye and many more with a simple pair of binoculars. This is probably the most famous of the amateur space-watchers' catalogues.

NGC is the *New General Catalogue*. This was published over a hundred years after the Messier Catalogue by Johan Ludvig Emil Dreyer (1852–1926) and contains thousands of objects, some extremely faint. The NGC includes all the Messier objects, so M42, the Orion nebula, is also NGC 1974. There is an extension to the NGC called the **IC** or *Index Catalogue*.

Mel is the *Melotte Catalogue* of star clusters compiled by Philibert Jacques Melotte (1880–1961). It was published in 1915 and contains 250 galactic star clusters. An accomplished observer, Melotte discovered Pasiphae, the eighth moon (at the time) of Jupiter, in February 1908 from the Royal Greenwich Observatory in London.

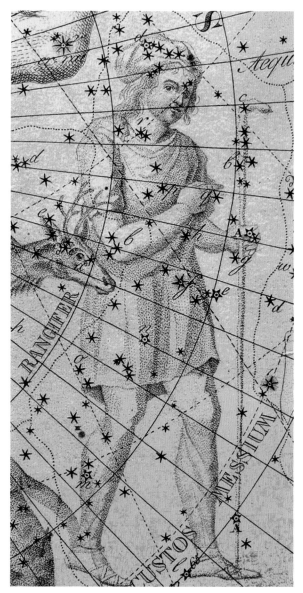

In honour of Charles Messier, the astronomer Joseph Lalande created the constellation of Custos Messium in the heavens. On the left a curious reindeer tries to get in on the scene, but unfortunately it and Messium had but a short time to ponder their circumpolar travels before history consigned them to the 'obsolete constellations' box.

DEEP-SKY OBJECTS

Here is the full list of deep-sky objects found on the charts and visible with just the eye:

Constellation	Deep-sky designation	Type	Magnitude	Size	Distance in light-years
Andromeda	M31	Galaxy	4.8	3°	~2.8 million
Coma Berenices	Mel 111	Galactic cluster	2.7	4' 30'	265
Cancer	M44	Galactic cluster	3.7	1° 35'	577
Canis Major	M41	Galactic cluster	4.5	38'	2,300
Canis Major	Mel 65	Galactic cluster	4.1	8'	5,000
Carina	Mel 82	Galactic cluster	3.8	30'	1,300
Carina	NGC 3114	Galactic cluster	4.2	35'	3,000
Carina	IC 2581	Galactic cluster	4.3	8'	2,868
Carina	IC 2602	Galactic cluster	1.9	50'	479
Carina	Mel 103	Galactic cluster	3.0	55'	1,300
Carina	NGC 3372	Nebula	5.0	2°	10,000
Centaurus	NGC 5139	Globular cluster	3.6	36'	17,000
Centaurus	NGC 3766	Galactic cluster	5.3	12'	5,500
Crux	NGC 4755	Galactic cluster	4.2	10'	7,600
Cygnus	M39	Galactic cluster	4.6	32'	825
Cygnus	NGC 7000	Nebula	-	2°	1,600
Dorado	LMC	Galaxy	0.4	9° 10' x 2° 50'	179,000
Dorado	NGC 2070	Nebula	5.0	40' x 20'	179,000
Gemini	M35	Galactic cluster	5.3	28'	2,800
Hercules	M13	Globular cluster	5.7	23'	25,300
Orion	M42	Nebula	4.0	1°	1,600
Perseus	NGC 869 & 884	Galactic clusters	4.7	1°	7,100
Perseus	NGC 1499	Nebula	5.0	2° 40' x 40'	1,000
Perseus	M34	Galactic cluster	5.2	35'	1,400
Perseus	Mel 20	Galactic cluster	2.9	3°	600
Puppis	M47	Galactic cluster	4.4	30'	1,600
Puppis	NGC 2451	Galactic cluster	2.8	50'	850
Sagittarius	M8	Nebula	5.8	1° 30 x 40'	5,200
Sagittarius	M22	Globular cluster	5.1	24'	10,000
Sagittarius	M24	Star clouds	4.5	1° 30'	10,000
Scorpius	M6	Galactic cluster	4.2	20'	2,000
Scorpius	M7	Galactic cluster	3.3	1° 20'	800
Scorpius	NGC 6231	Galactic cluster	2.5	15'	5,900
Scutum	M11	Galactic cluster	5.8	14'	6,000
Taurus	M45	Galactic cluster	1.5	1° 50'	380
Triangulum	M33	Galaxy	5.7	1°	~3 million
Tucana	SMC^2	Irregular galaxy	2.3	5° 19' x 3° 25'	196,000
Tucana	NGC 104	Globular cluster	4.0	30'	13,400
Vela	IC 2391	Galactic cluster	2.5	50'	580
Vela	NGC 2547	Galactic cluster	4.7	20'	1,950

Anton is the *Anton Vamplew Catalogue* (yes indeedy) of eight 'lost' constellations with a hint of their stories, told as we journey through the starry skies. The majority of these fanciful designs were from the 'crazy constellation creation period' of about 1750–1800, when lots of astronomers were naming groups of stars and putting them into their star atlases pretty much at random. Over time these new constellations were either forgotten or just accepted by subsequent astro-designers. So, over the following pages, you can see whether you think those in the Anton Catalogue deserved to be written off.

THE ANTON CATALOGUE

Everyone else seems to have one, so here's mine. It's a collection of things easy to overlook and for the most part sights that have been lost in the mists of time.

Catalogue number	Constellation	Type	Name
Anton 0	Vulpecula	Galactic cluster	The Coathanger
Anton 1	Ophiuchus	Old constellation	Taurus Poniatovii
Anton 2	Triangulum	Old constellation	Triangulum Minor
Anton 3	Aries	Old constellation	Musca Borealis
Anton 4	Gemini/Auriga	Old constellation	Telescopium Herschelii
Anton 5	Boötes	Old constellation	Quadrans Muralis
Anton 6	Eridanus	Old constellation	Sceptrum Brandenburgicum
Anton 7	Eridanus	Old constellation	Psalterium Georgii
Anton 8	Sagittarius	Old constellation	Teabagus

The great thing about the night sky is that there is so much to see with the unaided eye. There's no need for telescopes or even binoculars, and, in fact, I would recommend getting to know your way around the starry skies first before you started peering any deeper. Your friends will be more impressed with you pointing out the star Asellus Borealis (in Cancer), or saying, 'In between those two stars is the great nebula Bonanza' if you don't have a telescope to hand.

Here come the constellations. Take your time, take a deep breath and remember: Rome wasn't built in a day. Don't expect to learn everything at once – the starry skies are there to be enjoyed and slowly savoured.

OLD NAMES FOR THE CONSTELLATIONS

Throughout history, not only have constellations been rearranged or abandoned, they have also changed some of their names. Here is a list of present-day groups with their former, usually more ornate, names.

Present Latin name	English name	Original Latin name	Original English name	Designer
Antlia	The Air Pump	Antlia Pneumatica	The Air Pump	Nicholas La Caille
Avis	The Bird	Avis Indica	The Indian Bird	Keyser & de Houtman
Columba	The Dove	Columba Noae	Noah's Dove	Petrus Plancius
Fornax	The Furnace	Fornax Chemica	The Chemical Furnace	Nicholas La Caille
Mensa	The Table	Mons Mensae	Table Mountain	Nicholas La Caille
Norma	The Square	Quadra Euclidid	Euclid's Square	Nicholas La Caille
Octans	The Octant	Octans Hadleianus	Hadley's Octant	Nicholas La Caille
Pictor	The Painter	Equuleus Pictor	The Painter's Easel	Nicholas La Caille
Pyxis	The Compass	Pyxis Nautica	The Sailor's Compass	Nicholas La Caille
Reticulum	The Net	Reticulum Rhomboidalis	The Rhomboidal Net	Isaak Habrecht
Sculptor	The Sculptor	Apparatus Sculptoris	The Sculptor's Apparatus	Nicholas La Caille
Scutum	The Shield	Scutum Sobiescianum	Sobieski's Shield	Johann Hevelius
Sextans	The Sextant	Sextans Uraniae	Urania's Sextant	Johann Hevelius
Volans	The Flying Fish	Piscis Volans	The Flying Fish	Keyser & de Houtman
Vulpecula	The Fox	Vulpecula cum Ansere	The Fox & Goose	Johann Hevelius

An alternative name for Fornax Chemica shows how in 'the good ol' days' names were much more flexible and easy going. Plus, what a great design!

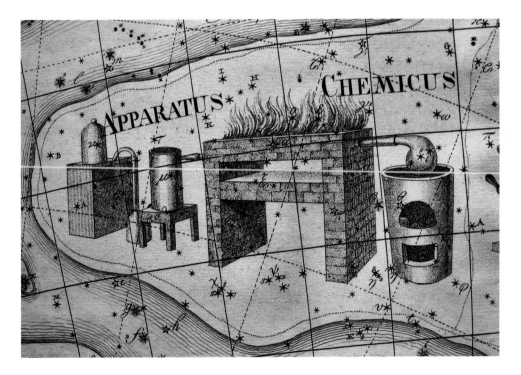

Here are the starry skies from the area around the North Celestial Pole.

The Northern Charts

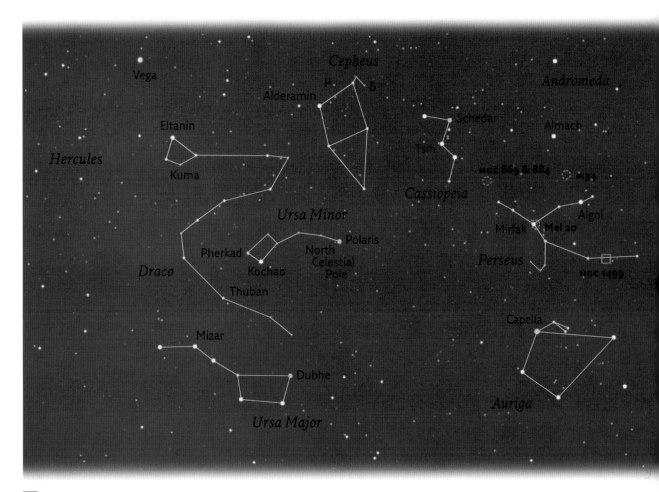

Galactic Cluster
Globular Cluster
Nebula
Galaxy

North Celestial Pole Chart

If you've been reading carefully up to this point then you will know all about the Plough and the starry movement around Polaris (the North Star). So if it is dark and clear right now, why not get a feel for the orientation of everything (all to do with the time and the direction you're looking) and head out for adventureland. On the chart you'll probably notice a rather large and empty area of the sky to the lower-right of Polaris, where nothing much seems to be happening. Actually most of this region is taken up with the wonderfully named constellation of Camelopardalis, the Giraffe, but it really is disappointingly faint.

January to March Skies

Winter: what a truly bright, twinkly, sparkly time of year this can be. The constellation of Orion stands out easily, surrounded by a plethora, nay, a cornucopia of stellar marvels. Bright stars have the ability to fire up the bits of your eye that can detect colour, making this a great time of year to find a number of tinted stars: overhead is yellow Capella, while high in the south is red Aldebaran. Orion itself gives you red Betelgeuse and bluish Rigel. As you gaze skyward you can ponder on how the Greeks could possibly have created the constellation Canis Minor, the Little Dog, out of just two stars. Perhaps the group was based on a small dog that met a Greek chariot!

Star Sights

The Orion nebula, M42
The Pleiades, M45, in Taurus
The Hyades star cluster in Taurus
Quadrantids meteor shower (peak around 3 January)

Galactic Cluster
Globular Cluster
Nebula
Galaxy

The northern winter skies looking south.

Orion

LATIN NAME
Orion
ENGLISH NAME
The Hunter
ABBREVIATION
Ori
LATIN
POSSESSIVE
Orionis

α STAR
Betelgeuse
MAGNITUDE
RANGE
0.3 to 1.0
STAR COLOUR
Red

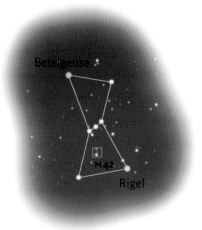

This is the brightest of all the constellations simply because it contains more brighter stars than any other, and so it dominates the winter skies. It is an ancient constellation with many stories attached to it, including that of the Scorpion (Scorpius) who was sent to kill Orion, which is why they were eventually placed on opposite sides of the sky.

Rigel has a bluish-white hue and is in fact brighter for most of the time than the (much mispronounced) leading star Betelgeuse (I like 'Beetle-Juice' – who wouldn't?). Betelgeuse is a truly massive variable star (see the star colours diagram on page 27), changing in brightness over a period of about six years.

In between Rigel and Betelgeuse you'll see three stars in almost perfect alignment which make up Orion's Belt. However, they are not actually linked in any way, and such an easily recognisable pattern is called an **asterism**. These stars are, from left to right, Alnitak, Alnilam and Mintaka.

LOCATION: Orion can be found wielding his club in the middle of the northern chart.

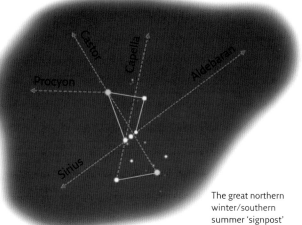

The great northern winter/southern summer 'signpost' Orion points the way to many nearby bright stars.

The wonderful Orion nebula as drawn by me on 10 January 1981 using a small 60 mm refractor. Larger telescopes will show more and more structure in this ginormous glowing cloud of gas.

DEEP-SKY OBJECT
M42
TYPE
Nebula
MAGNITUDE
4.0
SIZE
1°
DISTANCE IN
LIGHT-YEARS
1,600

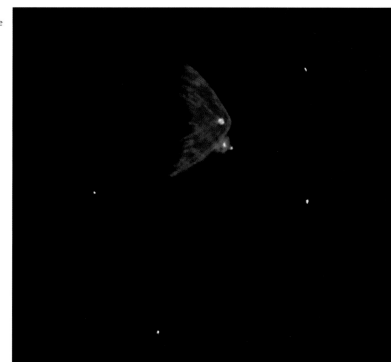

The Orion nebula is the famous faint fuzzy patch you can see with your unaided eye that sits directly below the three 'belt' stars. Also known as the Sword of Orion, this is an emission nebula glowing because the stars inside it (most notably θ Ori) are 'exciting' all the gases. Currently around 1,000 stars are being born here – a truly stellar nursery.

ORION NEBULA
NATIONAL PARK
1,600 LIGHT-YEARS

This magical nebula will undoubtedly become one of the top space-tourist visits in the future, andis seen in glorious colour on page 33.

Here's the classic Hunter constellation figure. Orion's lion shield can be fairly well seen in the night sky as a long curved grouping of six stars. Interestingly, all of these stars are catalogued with the Greek symbol π – from the top down we find π1 to π6.

Taurus

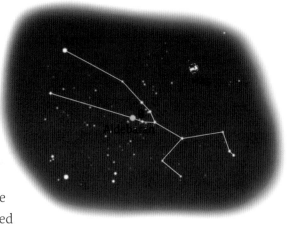

LATIN NAME
Taurus
ENGLISH NAME
The Bull
ABBREVIATION
Tau
LATIN
POSSESSIVE
Tauri

α STAR
Aldebaran
MAGNITUDE
0.85
STAR COLOUR
Orange

Taurus is an extremely ancient group – possibly one of the very first constellations designed. To the Egyptians it was the bull-god Osiris, while the Greeks built one of their usual myths around it: apparently it was Zeus who placed the Bull in the sky after it had safely carried one of his loves, the fair maiden Europa, to the island of Crete. If you look at the actual design it is only ever drawn as the front half of a Bull. This is easily explained because Taurus apparently swam all the way to Crete, so his back half was of course under water and invisible. There's an answer for everything.

It's interesting that early civilisations which were not connected to each other in any way also created the same animal in the sky. For example, Amazonian tribes depicted the V-shaped Hyades cluster, just as the Greeks did, as the head of a bull.

One of the seasonal jewels of the night sky is the red-tinted main star Aldebaran ('Follower'), which ranks as the 14th brightest star of the heavens.

LOCATION: Taurus is up to the right of Orion on the northern chart.

DOUBLE STAR
θ Tau
MAGNITUDE
3.4 & 3.9
SEPARATION
5' 37"
COLOURS
White and yellow

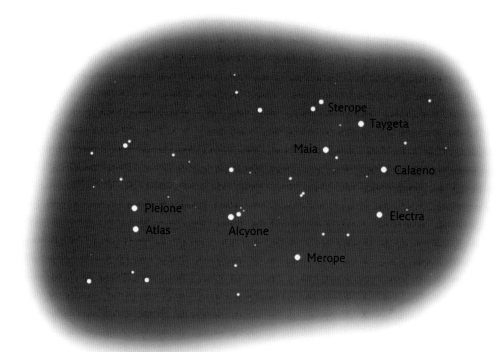

DEEP-SKY OBJECT
M45

TYPE
Galactic cluster

MAGNITUDE
1.5

SIZE
1° 50'

DISTANCE IN
LIGHT-YEARS
380

The Pleiades are one of the jewels of the skies. How many can you see with the unaided eye? I'm sure I've seen ten with my super vision, and that wasn't from a really dark location. If you count more than 30 you're probably already used to the phrase 'and I'm the Queen of Sheba'. The group actually contains many hundreds of stars, and binoculars or a telescope at very low power will show its full glory. The Pleiades are moving through a nebulous cloud, which shines by reflecting the light of the stars – but this only shows up on photographs.

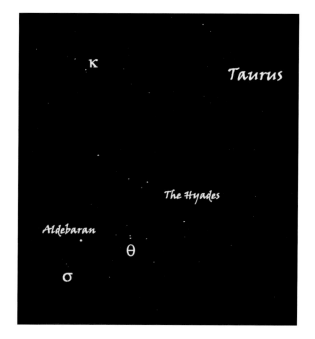

Some marvellous unaided-eye double stars show up well in this picture I took of Taurus. Aldebaran represents the red eye of the Bull, but visually forms part of the recognisably V-shaped Hyades star cluster.

Auriga

LATIN NAME
Auriga
ENGLISH NAME
The Charioteer
ABBREVIATION
Aur
LATIN
POSSESSIVE
Aurigae

α STAR
Capella
MAGNITUDE
0.08
STAR COLOUR
Yellow

In ancient times the Greeks identified Capella as Amalthea, who seemed to be either a beautiful young princess or a goat. Bad eyesight? Taking the goat's side, the story is that she helped feed the infant Zeus. However, one day, clumsy fool that Zeus was, he accidentally broke off one of her horns. Always looking for that happy ending, he wove his godly magic and made it a 'cornucopia', literally a 'horn of plenty' that would fill up with whatever its user wanted: crisps, nuts, profiteroles, water, tea, coffee, etc. Now I could do with one of those!

Capella is easily **circumpolar** from latitudes higher than 50°N, such as the UK, Vancouver, Canada, and Frankfurt, Germany. Actually here's another double star composed of two large stars orbiting each other, although you'd need an astronomically enormous telescope to see them both – so big, you would probably have a car sticker saying, 'My other telescope is in orbit!'

LOCATION: The Charioteer is at the very middle top of the northern chart, and to the lower right of the North Celestial Pole chart (see page 43).

While Auriga looks concerned that his chariot is missing, the goat myth is confirmed in this design with Capella ('the little she-goat') nestling comfortably in the arms of the Charioteer. Notice the small goats in his left hand, which lead to the two nearby stars being known as 'the kids'.

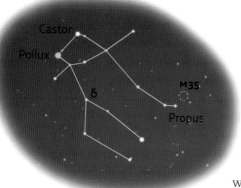

Gemini

Another one of the fine winter constellations, the two leading stars are the twins Castor ('the Warrior') and Pollux ('the Boxer'), who were Argonauts with Jason on the search for the Golden Fleece. Oddly, Pollux (β Gem) is brighter than Castor (α), but according to legend Castor has faded over the centuries.

Castor is really a double if you see it through a telescope. However, even then all is not as it seems – there are doubles galore in the Castor system. A total of six stars (three doubles) are orbiting around each other in times varying from nine days to 10,000 years!

δ Gem is an ordinary magnitude 3.5 white star. What gives it more interest, purely from a historical point of view, is that this is the location where Pluto was discovered in 1930.

LOCATION: The Twins are to the upper left on the northern chart.

LATIN NAME	Gemini
ENGLISH NAME	The Twins
ABBREVIATION	Gem
LATIN POSSESSIVE	Geminorium

α STAR	Castor
MAGNITUDE	1.58
STAR COLOUR	White

DEEP-SKY OBJECT	**M35**
TYPE	Galactic cluster
MAGNITUDE	5.3
SIZE	28'
DISTANCE IN LIGHT-YEARS	2,800

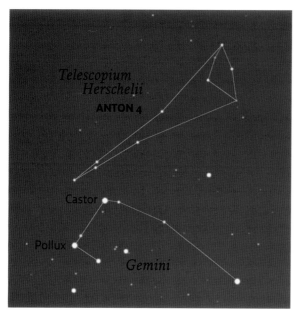

OBJECT	**ANTON 4**
TYPE	Old Constellation

Telescopium Herschelii, or Herschel's Telescope, was a creation of Abbé Hell in 1781. He was commemorating the recent discovery of the planet Uranus by William Herschel (the planet had initially been called Herschel).

This cluster of around 200 stars may just be seen with the unaided eye on super-clear nights.

April to June Skies

During spring one of the constellations that has had a lot of press to date is moving overhead: Ursa Major and its associated asterism, the Plough. I have a great affinity for this time of year as 'when I were a lad' these were the constellations that I learned first. Star shapes such as the great 'kite' of Boötes and the 'backwards question mark' of Leo are easy to find, and you can easily spot a sprinkling of the bright stars across the sky: Arcturus and Spica in the southeast, Regulus off to the right, with Castor and Pollux (leftovers from the winter) over in the west.

Star Sights

Using the Plough to find Arcturus and Capella
The Mizar-Alcor double star in Ursa Major
Mel 111 star cluster in Coma Berenices
The Beehive or Praesepe star cluster, M44, in Cancer
Lyrids meteor shower (peak around 22 April –
  see page 149)

- Galactic Cluster
- Globular Cluster
- Nebula
- Galaxy

The northern spring
skies looking south.

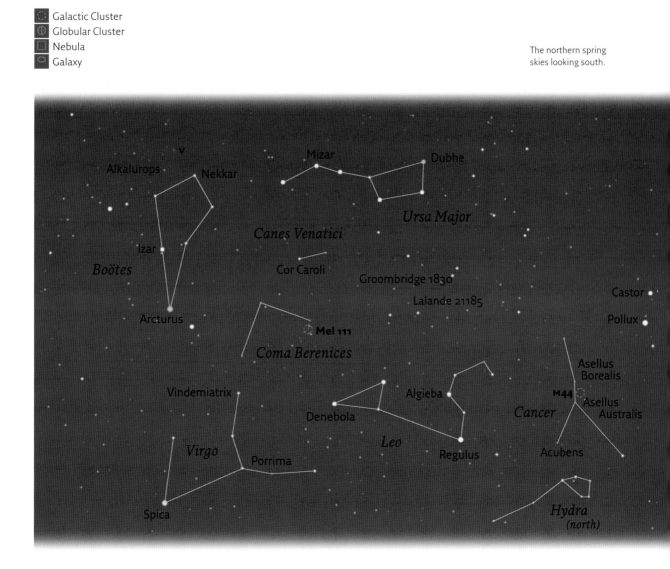

Ursa Major

LATIN NAME
Ursa Major
ENGLISH NAME
The Great Bear
ABBREVIATION
UMa
LATIN
POSSESSIVE
Ursae Majoris

α STAR
Dubhe
MAGNITUDE
1.79
STAR COLOUR
Orange

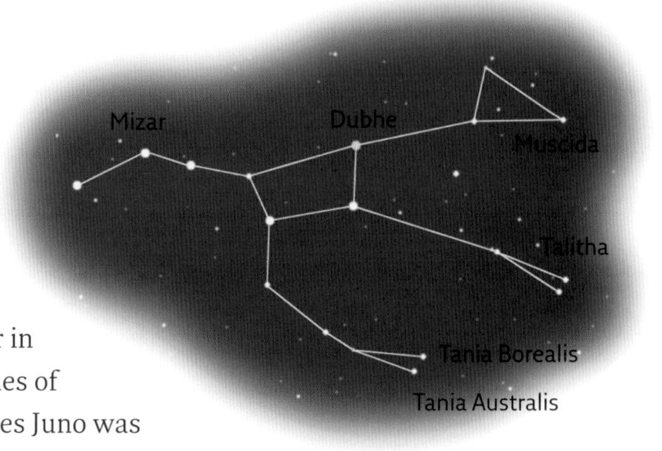

Originally the beautiful maiden Callisto who was changed into a bear by Jupiter in order to escape the evil clutches of her boss, Juno. In Roman times Juno was head of space, the Universe and everything, but she had a bit of a temper. The moral: having everything doesn't necessarily make you a nice person – an important lesson for life.

As has been mentioned, the most famous part of Ursa Major is a group of seven stars known in the UK as the Plough. However, due to its easily identifiable shape, it is known by many different names around the world: in Hindu astronomy we find Sapta Rishi, the Seven Sages, while in Mandarin it is Bei Dou Qi Xing, the Northern Ladle of Seven Stars.

Ursa Major has a few stars with fascinating names that happen to be moving all over the place: Lalande 21185 is magnitude 7.5 and moves by 4.8" every year.

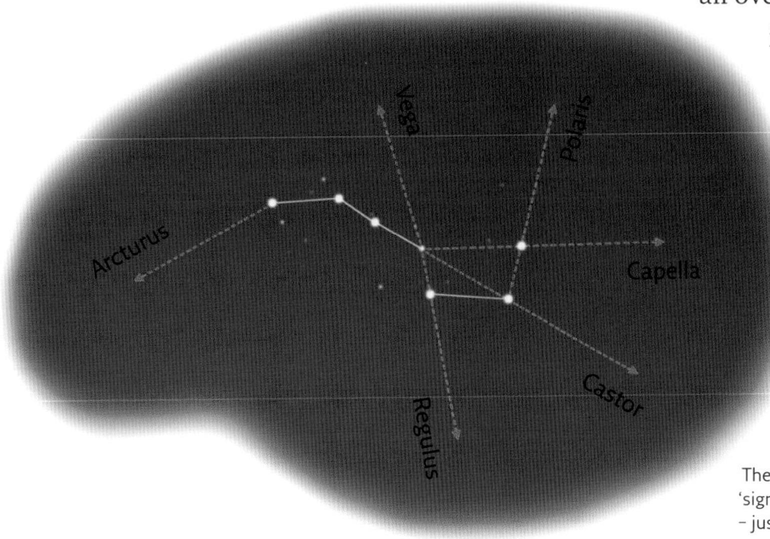

The Plough – a great 'signpost' to the stars – just look where you can get to!

Ursa Major is not bad for those Greek designers of Ye Olden Dayes – I mean it actually sort of looks like a bear once you've joined up the stars in a more or less bear-like fashion. Look closely and you'll see the seven famous stars of the Plough within the bear's body and tail.

It is only 8.3 light-years away and may have its own solar system of planets. Then there's Groombridge 1830, a star that's 29 light-years away. Shining at magnitude 6.4, it is moving across the sky at a zippy 7" every year. When everything is taken into account, Groombridge 1830 is moving through space at nearly 350 km per second! Regrettably neither of these stars is an unaided-eye object, but you can still gaze and wonder and point in their general direction.

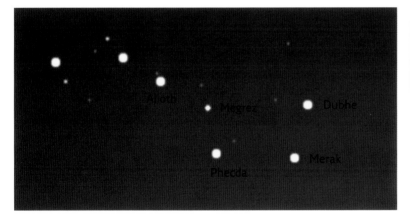

If you really want to get into the stargazing mood then remember that each star in the Plough has a name. Learn these now if you're keen to impress.

These two form the famous optical double star in the bend of the Plough's handle. If you have a telescope you'll see that Mizar is itself a double star, this time a true binary, with a companion 14.4" away. The other unlabelled star in the picture is not part of the group, but does have the great name of Sidus Ludovicianum.

DOUBLE STAR
Mizar and Alcor
ξ and 80 UMa

MAGNITUDE
2.2 & 4.0

SEPARATION
11' 49"

COLOURS
Both white

Ursa Minor

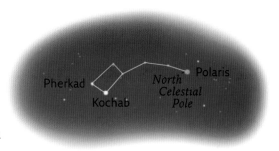

Pherkad

Kochab

North Celestial Pole

Polaris

LATIN NAME
Ursa Minor
ENGLISH NAME
The Little Bear
ABBREVIATION
UMi
LATIN POSSESSIVE
Ursae Minoris

α STAR
Polaris
MAGNITUDE
2.02
STAR COLOUR
Yellow

Ursa Minor was designed by the first great Greek astronomer, Thales, around 600 BC. It represents Arcas, son of Callisto – the maiden of Ursa Major fame. The main stars form a smaller version of the Plough, but in this case the handle is more curved. Some unknowing humans have confused the Plough and Ursa Minor for this reason – but after reading this, you have no excuse. Kochab (β) and Pherkad (γ) are known as the Guard Stars, for they are the guardians of the pole.

Polaris (α) has been a well-known star for a very long time. Of course we know it as the North Star and the Pole Star, but the early Greeks knew it as Phoenice ('lovely northern light'), the Anglo-Saxons as Scip-steorra ('the ship star'), and early British mariners as the Steering Star. The list goes on and on, which shows the importance of this star in history.

LOCATION: The Little Bear prowls around the centre of the North Celestial Pole chart (see page 43).

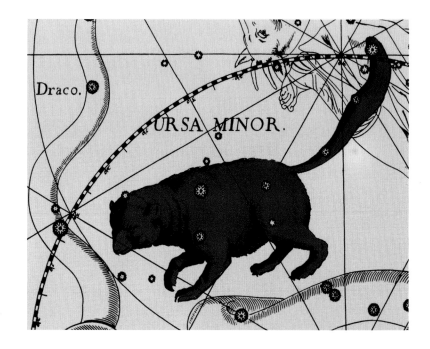

The Little Bear as depicted in the Hevelius star atlas *Uranographia*

Draco

This ancient group was probably based on the dragon Ladon, the guardian of the golden apples in the Garden of Hesperides. Draco sits in the heavens between Ursa Major and Ursa Minor; it's fairly dim but easily identifiable. You can weave with the stars from the Dragon's tail, above the Plough, along the body to the fire-breathing head.

Draco used to be famous about 4,000 years ago when Thuban was the Pole Star (that is, the star closest to the North Celestial Pole). Now it is a lesser celebrity.

LOCATION: Draco smoulders on the North Celestial Pole chart (see page 43).

(see page 43)

LATIN NAME
Draco
ENGLISH NAME
The Dragon
ABBREVIATION
Dra
LATIN POSSESSIVE
Draconis

α STAR
Thuban
MAGNITUDE
3·7
STAR COLOUR
White

Camelopardalis

Notice the North Celestial Pole Chart has a great big empty(ish) patch. Well that's where our friendly Giraffe is. It was designed in 1614 by Jacobus Bartschius as a camel who eventually went through a transformation to become a Giraffe. There is nothing of note for the eye and even your dot-to-dot imagination will be hard pressed to create anything at all. It's a shame as the constellation's name is fantastic – put the stress on 'par'.

To make matters worse even its leading star has no name: it's as though Jacobus was a dodgy builder who left before completing the job. Maybe if he had known one super fact he would have been a little more considerate. This fact is: α Cam is around an incredible 6,900 light-years distant (give or take 3,000!), probably making it one of the furthest stars you can see with the unaided eye.

LATIN NAME
Camelopardalis
ENGLISH NAME
The Giraffe
ABBREVIATION
Cam
LATIN POSSESSIVE
Camelopardalis

α STAR
–
MAGNITUDE
4·3
STAR COLOUR
Blue

Some nearby stars have been marked for your convenience, otherwise it's not that easy to locate the faint Giraffe.

Boötes

LATIN NAME
Boötes
ENGLISH NAME
The Herdsman
ABBREVIATION
Boo
LATIN
POSSESSIVE
Boötis

α STAR
Arcturus
MAGNITUDE
-0.04
STAR COLOUR
gold

Now this is the Herdsman, or Ploughman, who drives the Bear (Ursa Major) around the sky while holding on tightly to his hunting dogs (Canes Venatici). Honestly, either I'm missing something or this Herdsman has a problem, as his shape eludes me totally, even in my most creative moments. If you can join these stars up to make anything like a man with a herd I want to hear from you. To me the bright main stars in the northern hemisphere form an upside-down teardrop or an elongated kite shape with the bright golden Arcturus at its bottom tip.

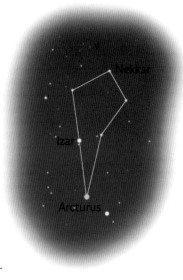

A meteor shower (see page 146 for an explanation of what that is) from Boötes called the June Boötids also takes the superb name of the pons-Winneckids.

LOCATION: The Herdsman is herding on the left side of the northern chart.

DOUBLE STAR
v Boo
MAGNITUDE
5.0
SEPARATION
10' 28"
COLOURS
Deep orange and white

Arcturus can easily be found by following the handle of the Plough down in a gently curving arc. Plus did you know that Arcturus, Greek for the Bear Keeper, is a roving star? Its movement over 1000 years is about the same as the width of the Full Moon. This is because it is a relatively close star, only 37 light-years away. Arcturus is also the fourth brightest star in the sky – so, all in all, there's plenty to talk about.

DEEP-SKY OBJECT
ANTON 5
TYPE
Old Constellation

In this faint (and I mean faint) corner of the sky there used to be the old constellation of Quadrans Muralis, the Mural Quadrant, created by Joseph Lalande in 1795. Even though Quadrans no longer exists, it is remembered because of the meteor shower that appears from this area at the start of each year – the Quadrantids.

Canes Venatici

Polish astronomer Johann Hevelius added this faint constellation to the sky in 1690 when nobody was looking. Sitting just underneath the Plough it represents Chara and Asterion, the two hunting dogs of Boötes, the Herdsman, who are out for a spot of bear worrying.

Comet hero Edmund Halley named the leading star Cor Caroli, which means 'Charles' heart', after King Charles II. Apparently this was because the star had shone brilliantly on the evening before the King's return to London on 29 May 1660, after years of exile.

LOCATION: Underneath the handle of the Plough is where you will find these hunting dogs.

LATIN NAME
Canes Venatici
ENGLISH NAME
The Hunting Dogs
ABBREVIATION
CVn
LATIN POSSESSIVE
Canum Venaticorum

α STAR
Cor Caroli
MAGNITUDE
2.9
STAR COLOUR
White

VARIABLE STAR
La Superba
Y CVn
MAGNITUDE RANGE
5.2 to 10
PERIOD
251 days

This star was named 'The Superb' by the 19th-century Italian Father Secchi because of what he thought was its amazing red colouring. Note that the star goes below the unaided-eye limit of magnitude 6, so it would be interesting to see at what magnitude it actually disappears and reappears from your location (that's if you can see it at all!); this will be an indication of how clear your skies are.

Canes Venatici, the Hunting Dogs, a design by Johann Hevelius as shown in his *Uranographia* star atlas.

DEEP-SKY OBJECT
Mel 111
TYPE
Galactic cluster
MAGNITUDE
2.7
SIZE
4' 30"
DISTANCE IN LIGHT-YEARS
265

Coma Berenices

Let's all throw our hair into the sky! That's what happened to the wonderful locks of Queen Berenice. She was the wife of Ptolemy III, the King of Egypt. After a successful battle the goddess Aphrodite decided the sky would be a good place for her hair. And why not? Apparently she had dark hair; that's why it's all very faint. An old tale this may be, but Coma Berenices didn't become a fixed constellation until catalogued by Tycho Brahe in 1601.

Those with telescopes can find galaxies galore scattered like dandruff through this region of the sky.

LOCATION: You'll find Coma Berenices to the right of the bright golden star Arcturus on the northern chart.

The Coma star cluster of about 45 stars used to be the fuzzy hair at the end of Leo the Lion's tail; now it forms the flowing hair of Queen Berenice.

LATIN NAME
Coma Berenices
ENGLISH NAME
Berenice's Hair
ABBREVIATION
Com
LATIN POSSESSIVE
Comae Berenices

α STAR
Diadem
MAGNITUDE
4.3
STAR COLOUR
Yellow

Leo

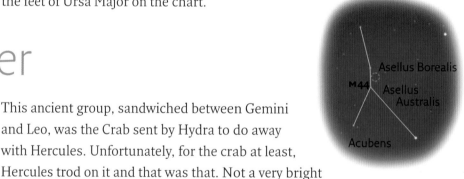

LATIN NAME
Leo
ENGLISH NAME
The Lion
ABBREVIATION
Leo
LATIN
POSSESSIVE
Leonis

α STAR
Regulus
MAGNITUDE
1.35
STAR COLOUR
White

Leo is one of the ancient designs in Greek and Roman legend representing the lion happily wandering around in the Nemean forest. Along comes Hercules on his 12 labours quest and strangles our feline friend – the Classical tales are like that. On a visual note, unlike many other constellations, Leo could be said to resemble what it is supposed to be: the Lion's head is a large backward question mark with its body off to the left.

Regulus, at the base of the head, is very close to the **ecliptic** and as a result is one of only four bright stars that can be covered by the Moon and the planets. The astro-technical term for this is an **occultation**.

LOCATION: Leo forms a distinctive shape underneath the feet of Ursa Major on the chart.

Cancer

LATIN NAME
Cancer
ENGLISH NAME
The Crab
ABBREVIATION
Cnc
LATIN
POSSESSIVE
Canceri

α STAR
Acubens
MAGNITUDE
4.25
STAR COLOUR
White

This ancient group, sandwiched between Gemini and Leo, was the Crab sent by Hydra to do away with Hercules. Unfortunately, for the crab at least, Hercules trod on it and that was that. Not a very bright constellation, it has to be said, but what it lacks in visual excitement it makes up for in the great Beehive star cluster.

LOCATION: The faint crab is found in the rocky space-pool to the right of Leo in the northern chart.

DEEP-SKY OBJECT
M44
TYPE
Galactic cluster
MAGNITUDE
3.7
SIZE
1° 35'
DISTANCE IN
LIGHT-YEARS
577

The Beehive or Praesepe cluster has hundreds of stars, many of them doubles, visible to the eye as a 'fuzzy' patch. Due to its brightness, M44 has been known since ancient times.

Detail of the Beehive seen through a good pair of binoculars

Virgo

An ancient group associated with the goddess of justice. Apparently, a little cheesed off about how humans were treating the Earth, she left her body to seek happiness in the stars, becoming the Maiden or Virgin (hence Virgo). Who can blame her? You may think that the second largest constellation in the sky would have heaps to offer us visually; alas, it doesn't, apart from the bright leading star Spica.

Vindemiatrix (ε Vir) ('the Grape Gatherer') is a star with a link to drink: the first rising of this star marked the start of the new wine vintage. Cheers!

LOCATION: Virgo is taking a rest on the lower left of the northern chart.

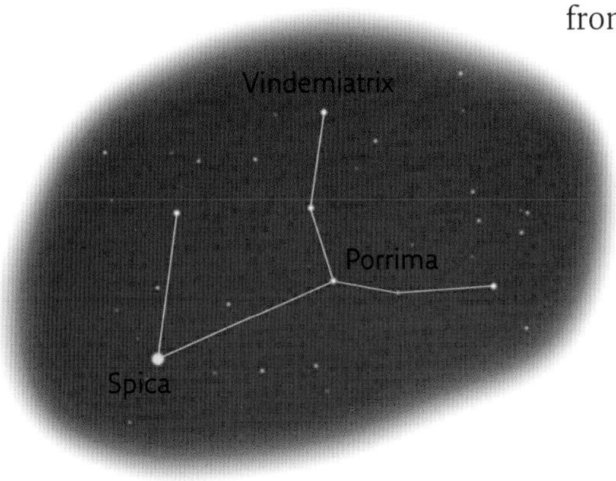

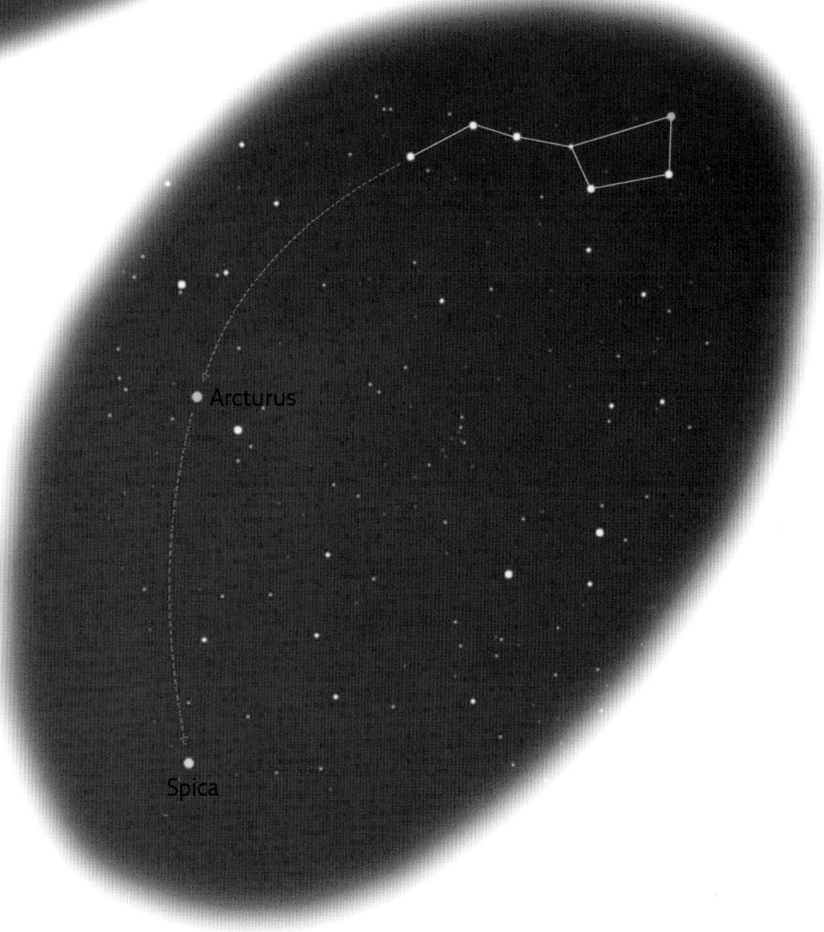

LATIN NAME
Virgo
ENGLISH NAME
The Maiden
ABBREVIATION
Vir
LATIN POSSESSIVE
Virginis

α STAR
Spica
MAGNITUDE
0.98
STAR COLOUR
Bluish-white

To find Spica, continue that arc that went from the Plough to Arcturus. Occasionally the planets or the Moon can cover, or occult, Spica because it is close to the ecliptic. The other bright stars this can happen to are Aldebaran (in Taurus), Regulus (in Leo) and Antares (in Scorpius).

July to September Skies

Over summer we begin to see the Milky Way appearing from the eastern horizon. Of course, in the northern hemisphere we have the problem of the Earth tilting towards the Sun, which gives us nice long warm summer days, but leaves the time for starry sky viewing very limited indeed – not that I'm complaining. Over in the west is a remnant of last season: the bright star Arcturus, leading the constellation of Boötes. To its left is the splendid curving constellation Corona Borealis, while a vast swathe in the south is dominated by the Serpent Bearer, Ophiuchus. Over in the east the Summer Triangle appears (which strangely is much more of an autumn group – ah, well).

Star Sights

The Great Cluster, M13, in Hercules
The abandoned constellation of Taurus Poniatovii
Perseids meteor shower (peak around 12 August)
Piscids meteor shower (two peaks, around 8 and
  21 September)

- Galactic Cluster
- Globular Cluster
- Nebula
- Galaxy

The northern
summer skies
looking south.

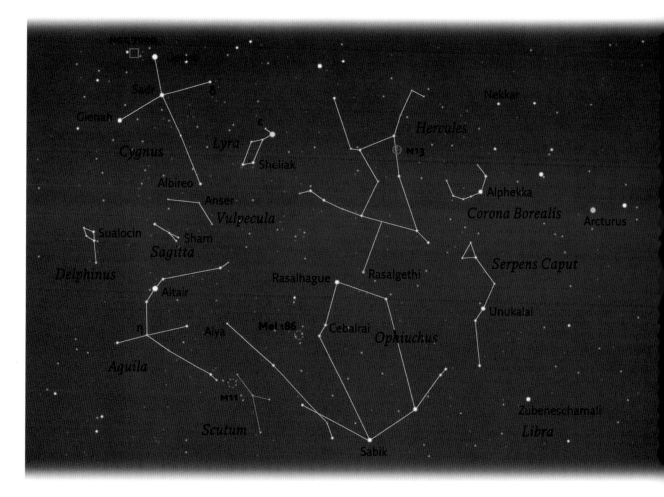

Cygnus

LATIN NAME
Cygnus
ENGLISH NAME
The Swan
ABBREVIATION
Cyg
LATIN POSSESSIVE
Cygni

α STAR
Deneb
MAGNITUDE
1.25
STAR COLOUR
White

Cygnus is one of the ancient constellations. According to one story, it represents the god Jupiter on a secret mission of love to Leda, wife of Tyndarus. He cleverly, or so he thought, changed himself into a swan so as to remain incognito. The fact that we know the story would indicate his evasive measures did not stand the test of time.

The Swan flies majestically along the Milky Way, the faint milky band that is in reality millions of distant stars that make up a part of our Galaxy. Away from light-polluted skies, you can see this river of haze at its finest; with binoculars it explodes into views of star clusters, nebulae and all manner of wonders.

Deneb is the star that marks the swan's tail, while Albireo is the head. Together with the outstretched wings, formed by Gienah through Sadr to δ Cyg, you can see why Cygnus is also called the Northern Cross.

LOCATION: The happy Swan is to the top left of the northern chart.

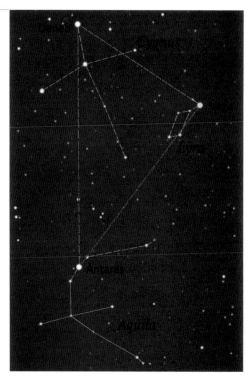

The bright stars of Deneb (~2,100 light-years away), Vega (25 light-years) and Altair (16 light-years) form the famous Summer Triangle.

DEEP-SKY OBJECT
M39
TYPE
Galactic cluster
MAGNITUDE
4.6
SIZE
32'
DISTANCE IN LIGHT-YEARS
825

Placed in the Messier Catalogue as late as 1764, this starry group is nevertheless so bright that Aristotle noted it in ancient Greek times. Although we are viewing the summer skies here, this cluster is shown in the top right of the northern October to December chart. Just one of those things, I guess.

DEEP-SKY OBJECT
NGC 7000
TYPE
Nebula
SIZE
2°
DISTANCE IN LIGHT-YEARS
1,600

The North America nebula can 'allegedly' be picked out (although I have never seen it, and many astronomers, most notably Greg, have a very strong word indeed for people who say they have!) on really dark nights away from light pollution. This cloudy patch sits in the heart of the Milky Way, taking its name from its obvious shape. So, can you see this nebula? Call Greg if you can.

Lyra

A constellation of antiquity showing the musical instrument invented by the messenger god Hermes and given to his half-brother Apollo, god of music.

Vega (α) is a relatively close (27 light-years-distant) star that held the post of Pole Star over 11,000 years ago – and it'll do the job once more in about 14,500 AD. This is due to the spinning wobble of the Earth that slowly moves our 23.5° axial tilt around in a 25,800-year circle. The point above the North or South Pole changes over the same period, hence the changes in pole stars. Vega is the third brightest star visible in the northern hemisphere, after Sirius and Arcturus, and in 1850 it became the first star ever to be photographed.

LOCATION: An excellent little constellation to the right of Cygnus on our chart, with its leading star Vega being the brightest of the Summer Triangle.

LATIN NAME
Lyra
ENGLISH NAME
The Harp
ABBREVIATION
Lyr
LATIN POSSESSIVE
Lyrae

α STAR
Vega
MAGNITUDE
0.03
STAR COLOUR
Bluish-white

DOUBLE STAR
η1 and η2 Lyr

Here's a double star worth trying to take a look at. Epsilon 1 and 2 are an optical double with a wide separation that is supposed to be a test of good eyesight. Now pick up your telescope (what do you mean, you haven't got one?) because ε Lyr is also a double-double – two stars that are themselves two stars.

A stellar bonanza indeed: four stars for the price of one!

MAGNITUDE
5.0 & 5.0
SEPARATION
3.5'
COLOURS
Yellow and orange

Vega (also Wega) was the 'Vulture Star' in ancient Egypt, and that is clearly shown in this 1801 design by Johann Bode.

VARIABLE STAR
Sheliak
β Lyr

MAGNITUDE RANGE
3.34 to 4.3
PERIOD
12.9 days

An eclipsing binary variable. This is the first of the beta Lyr type variables where the stars orbit so close together that gravity pulls them out of shape – making them look more egg-shaped than spherical!

Aquila

LATIN NAME
Aquila
ENGLISH NAME
The Eagle
ABBREVIATION
Aql
LATIN
POSSESSIVE
Aquilae

α STAR
Altair
MAGNITUDE
0.77
STAR COLOUR
White

This old constellation, representing the feathered friend of the god Jupiter, is often depicted carrying his thunderbolts – now there's a job. The good ol' Milky Way runs behind the Eagle, making this area well worth a meandering gaze in darker skies. As for the leading twinkler, Altair, it is only about 16 light-years away, which makes it one of the closest stars to us.

Above and to the left of Aquila is the small constellation of Delphinus, the Dolphin. Apart from its fine appearance, I mention this group because of its star names: alpha Delphini is Sualocin, while beta Delphini is Rotanev. Reverse the spellings and you get Nicolaus Venator, who was an assistant to the 17th- and 18th-century Italian astronomer Giuseppe Piazzi. Cheeky boy.

LOCATION: Altair is flying down on the lower left of the northern chart.

Aquila, the Eagle, in ye olden dayes used to carry Antinous on his journeys around the sky. As you can see, Antinous was not a small chap, and eventually the Eagle had enough and let go – Antinous dropped off the page never to reappear in the constellation club.

DOUBLE STAR
η Aql
MAGNITUDE RANGE
3.5 to 4.3
PERIOD
7.176 days

A Cepheid variable. You can use the nearby β Aql, which shines away at magnitude 3.7, as a useful comparison star.

Anser

Vulpecula

Hevelius created this small constellation as Vulpecula cum Ansere, the Fox and Goose. We've since lost the goose, presumably eaten by the fox. Whatever the story, it is not the most distinctive of groups, but there is a great object to find – see below.

LOCATION: The Fox lurks underneath Cygnus on our chart. After that goose I guess the swan looks pretty tasty too.

LATIN NAME	Vulpecula
ENGLISH NAME	The Fox
ABBREVIATION	Vul
LATIN POSSESSIVE	Vulpeculae
α STAR	Anser
MAGNITUDE	4.44
STAR COLOUR	Orange

The original full constellation from Hevelius' star atlas *Uranographia* shows both the Fox and Goose happily wandering the heavens.

DEEP-SKY OBJECT
ANTON 0

TYPE
Galactic cluster

MAGNITUDE
3.6

SIZE
1°

The Coathanger, also known as CR 399 or Brocchi's cluster, takes its name from the astrotastically fab shape of its ten main stars. It's a coathanger all right! Somewhere to hang your spacesuit when you've been out on a very long drive. You can just see it with the unaided eye as a fuzzy patch in a dark sky, but you should really view it through binoculars.

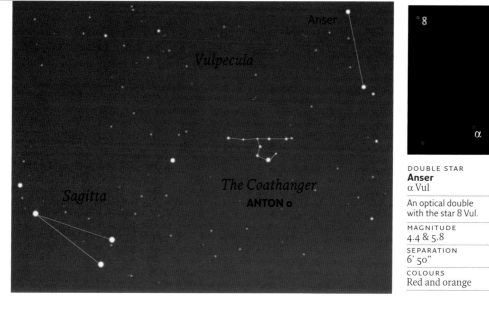

DOUBLE STAR
Anser
α Vul

An optical double with the star 8 Vul.

MAGNITUDE
4.4 & 5.8

SEPARATION
6' 50"

COLOURS
Red and orange

Hercules

LATIN NAME
Hercules
ENGLISH NAME
Hercules
ABBREVIATION
Her
LATIN POSSESSIVE
Herculis

α STAR
Rasalgethi
MAGNITUDE RANGE
3.1 to 3.9
STAR COLOUR
Red

Originally an ancient Greek constellation representing the strongest man in the world, who was placed in the sky after finishing 12 supposedly impossible 'labours'. He's a readily identifiable cheeky chap to the right of Cygnus, with four central stars making the easily remembered 'four-star-trapezial-rhomboidal-quadratic' shape.

LOCATION: With his join-the-dots shape looking like a hop-scotching Morris dancer, Hercules can be found just above the centre on the northern chart.

DEEP-SKY OBJECT
M13
TYPE
Globular cluster
MAGNITUDE
5.7
SIZE
23'
DISTANCE IN LIGHT-YEARS
25,300

The Great Cluster, as it is named, is the best globular cluster in the northern hemisphere. This is best observed from a dark site when its 10' size is fairly easy to see with the unaided eye. This round 'blob' looks great with binoculars, but you really need a telescope to start resolving the 'fuzz' into stars.

VARIABLE STAR
Rasalgethi
α Her
MAGNITUDE RANGE
2.8 to 4.0
PERIOD
~3 months

Not only variable, but also a double – some stars really do give value for money. Rasalgethi's companion has a magnitude of 5.4 with a 5" separation, which means a telescope is called for.

Ophiuchus

LATIN NAME
Ophiuchus
ENGLISH NAME
The Serpent Bearer
ABBREVIATION
Oph
LATIN POSSESSIVE
Ophiuchi

α STAR
Rasalhague
MAGNITUDE
2.08
STAR COLOUR
White

A Greek constellation whose identity and story have been somewhat lost in the mists of time. Maybe he is Aesculapius, the Greek god of medicine – you know, the one who carries a serpent in a stick wherever he goes. In his hands is this beast, stretching to the right as the constellation Serpens Caput, the head, and to the left as Serpens Cauda, the tail.

On some old star maps Ophiuchus is written as Serpentarius, which is quite a good name for a serpent bearer.

LOCATION: Our man grapples with the serpent in the lower middle of the chart.

OBJECT
ANTON 1

TYPE
Old constellation

Centred around the Mel 186 galactic cluster, here is the ex-constellation of Taurus Poniatovii, or Poniatowski's Bull. Created in 1777 by Abbé Poczobut to honour the King of Poland, Stanislaus Poniatovii, it was so named because the V-shaped star formation looks like a smaller, fainter version of the Hyades cluster in Taurus, the Bull. There's more! Here we also find Barnard's Star – see right.

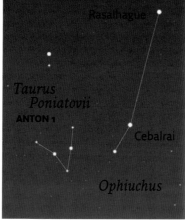

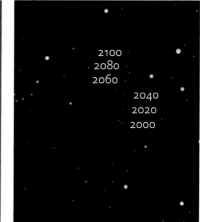

Barnard's Star (Velox Barnardi), named after its discoverer Edward Barnard, is the third closest star to us at just 6 light-years away. Barnard's Star really whizzes about the place – it has the largest motion of any star, moving at an incredible 10.25" per year. Unfortunately it is truly faint, and even its closeness only helps the magnitude reach 9.5.

Scutum

Formed by Hevelius is 1690 with the original name Scutum Sobiescianum, Sobieski's Shield, after Jan Sobieski, the King of Poland, for his heroic battles – oh, and he helped Hevelius after his observatory burned down.

Alpha Scuti doesn't actually have a name, so I have designated it 'Sobieski' in remembrance of the full title of the original constellation. It's been done before – just take a look at Vulpecula, the Fox, with its star Anser.

LOCATION: The Shield is protecting the bottom of the northern chart, underneath and to the right of Aquila.

LATIN NAME
Scutum
ENGLISH NAME
The Shield
ABBREVIATION
Sct
LATIN POSSESSIVE
Scuti

α STAR
Sobieski
MAGNITUDE
4.0
STAR COLOUR
Yellow

DEEP-SKY OBJECT
M11

TYPE
Galactic cluster

MAGNITUDE
5.8

SIZE
14'

DISTANCE IN LIGHT-YEARS
6,000

The Wild Duck cluster needs a really dark sky – so I hope your location is not too 'fowl'. Discovered in 1681 by Gottfried 'Donald' Kirch.

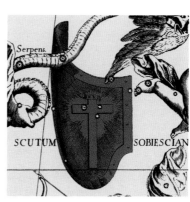

Scutum is seen here in all its glory, taken from *Uranographia*, the 1690 star atlas of Johann Hevelius.

October to December Skies

Some astronomy boffins take a holiday in autumn, claiming it's a quiet time for the constellations. It is true that a large area of the southern sky is filled with fainter stars, but overhead there's Cassiopeia, Perseus and the Milky Way. Also, in the south to west is the strangely named Summer Triangle – three stars that are best seen in the autumn! Okay, vast numbers of bright stars do not abound, but there are many abandoned constellations, an interesting variable star and some fine deep-sky objects to find. Plus, especially after the short summer nights, I'm sure it'll be a relief to get back out for some stargazing.

Star Sights

The Andromeda galaxy, M31
The Sword-Handle clusters in Perseus
The Summer Triangle
Square of Pegasus
Orionids meteor shower (peak around 21 October)
Geminids meteor shower (peak around 14 December)

Galactic Cluster
Globular Cluster
Nebula
Galaxy

The northern autumn
skies looking south.

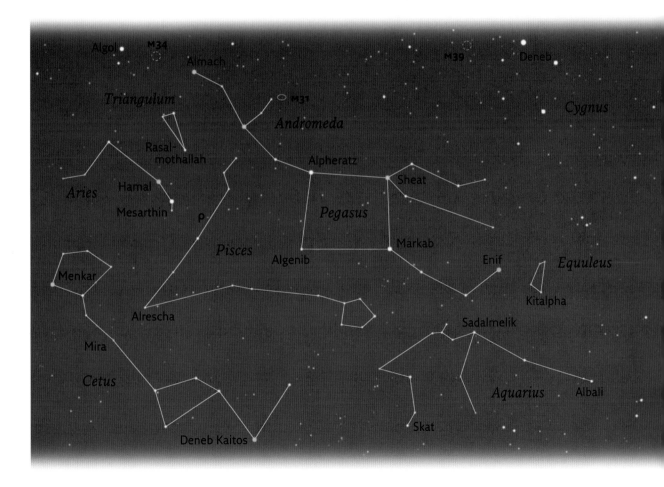

Cassiopeia

LATIN NAME
Cassiopeia
ENGLISH NAME
The Queen
ABBREVIATION
Cas
LATIN
POSSESSIVE
Cassiopeiae

α STAR
Schedar
MAGNITUDE
2.2
STAR COLOUR
Yellow

Cassiopeia was the outspoken Greek queen who caused her daughter Andromeda a lot of grief (see below). Her constellation is very easily found because its brightest stars form a 'W' shape in the heavens, and she is always visible from a large chunk of the northern hemisphere thanks to the fact that the stars sit quite close to the Pole Star.

The two stars Tsih (γ) and Schedar (α) can be used like the pointers of the Plough to find the Square of Pegasus.

LOCATION: Cassiopeia sits next to her husband, Cepheus, on the North Celestial Pole chart (see page 43).

VARIABLE STAR
γ Cas
MAGNITUDE RANGE
1.6 to 3.0
PERIOD
~0.7 days

This unstable star can change brightness quite quickly, so it's worth keeping an eye on it whenever you're out stargazing. This is an example of an irregular-period variable.

Perseus

LATIN NAME
Perseus
ENGLISH NAME
Perseus
ABBREVIATION
Per
LATIN
POSSESSIVE
Persei

α STAR
Mirfak
MAGNITUDE
1.8
STAR
COLOUR
yellow

Son of Zeus and Danaë, this was the chap who cut off the head of that snaky-haired Medusa, as marked by the star Algol ('El Ghoul'), and then flew with his winged plimsolls to save Andromeda from the Sea Monster – he was busy that day. The Milky Way runs right through Perseus, making it a fine view for those with darker skies.

LOCATION: You can find our hero Perseus on the North Celestial Pole chart (see page 43).

VARIABLE STAR
Algol
α Per
MAGNITUDE RANGE
2.1 to 3.4
PERIOD
2 days 20 hours 48 minutes

An eclipsing binary whose light levels can be observed changing over the course of an evening – which is why it also has the name of the 'Winking Demon'. The minimum magnitude lasts for ten hours. This was the second binary star discovered - in 1667 by Geminiano Montaniari.

DEEP-SKY OBJECT
NGC 869 and 884

TYPE
Galactic clusters

MAGNITUDE
4.3 and 4.4

SIZE
30' each

DISTANCE IN
LIGHT-YEARS
7,100 & 7,400

The so-called Sword-Handle is a wondrous double galactic cluster of magnitude 4.4 and 4.7 respectively. They are both ½° in diameter and together are certainly a great find. They are visible to the unaided eye, but a sweep of the area with binoculars is well worth while.

DEEP-SKY OBJECT
NGC 1499

TYPE
Nebula

MAGNITUDE
5.0

SIZE
2° 40' x 40'

DISTANCE IN
LIGHT-YEARS
1,000

The California nebula. Just visible with the eye, and, yes, with more power you can see that it is shaped like that west coast US state. This is located on the northern winter chart as well as on the North Celestial Pole chart.

DEEP-SKY OBJECT
M34

TYPE
Galactic cluster

MAGNITUDE
5.2

SIZE
35'

DISTANCE IN
LIGHT-YEARS
1,400

Tricky indeed to see. Dark skies a must.

DEEP-SKY OBJECT
Mel 20

TYPE
Galactic cluster

MAGNITUDE
2.9

SIZE
3°

DISTANCE IN
LIGHT-YEARS
600

α Perseus moving cluster. Sounds exciting, doesn't it? Does the name give away where it is? This is a sprawling family of stars based around the star α Per, which makes it very easy to find.

Cepheus

He's the King of Ethiopia, husband of Cassiopeia and father of Andromeda. Not incredibly bright (in stars, not as a king), but due to its proximity to the Milky Way there are quite a few galactic clusters and wispy nebulae to have a look at.

LOCATION: Cepheus waits patiently for a cup of tea towards the top of the North Celestial Pole chart (see page 43).

LATIN NAME
Cepheus
ENGLISH NAME
The King
ABBREVIATION
Cep
LATIN
POSSESSIVE
Cephei

α STAR
Alderamin
MAGNITUDE
2.44
STAR COLOUR
White

VARIABLE STAR
δ Cep

MAGNITUDE RANGE
3.9 to 5.0

PERIOD
5.3663 days

A most interesting yellow star in that it is variable with an exact period. This star was the first of a whole new class of variables named Cepheids (see page 30), stars whose magnitude fluctuates precisely with their period.

VARIABLE STAR
μ Cep

MAGNITUDE RANGE
3.43 to 5.1

PERIOD
~730 days

Named the Garnet Star by William Herschel due to its intense deep-red colour. This is a semi-regular variable type.

Pegasus

LATIN NAME
Pegasus

ENGLISH NAME
The Flying Horse

ABBREVIATION
Peg

LATIN
POSSESSIVE
Pegasi

α STAR
Markab

MAGNITUDE
2.49

STAR COLOUR
White

The Greek winged horse born from the blood of Medusa after Perseus chopped her head off. The square of the flying horse Pegasus is one of the landmarks in autumn skies. Now here's a strange thing – when is a square not a square? Enter the all-seeing body that looks after the sky and all that's in it – the International Astronomical Union, or IAU. They name asteroids, calculate orbits, and 'fix' general starry things. In 1923, in their wisdom and for no apparent reason, they took the top left star of the square, Alpheratz, and 'fixed' it in the constellation of Andromeda. The star to this day is α Andromedae (i.e. of the constellation of Andromeda) – it's just something we have to live with.

As an indication of how clear your skies are, see how many stars you can count inside the actual square. If the answer is none, then your skies are extremely unclear!

LOCATION: Our square horse flies right across the middle of the northern chart.

Pegasus is one of the few northern designs that appear upside-down as seen from the northern hemisphere. The smaller horse to the lower right is Equuleus – the constellation that sits in the next stable.

Equuleus

Kitalpha

A small, originally ancient Greek grouping with one story telling of this young horse being a present from Castor to Mercury – possibly for his birthday. Mercury apparently wanted to call it Trigger, but was advised against it.

This second smallest constellation is famed throughout history for the famous Great Daylight meteor showers of the early seventh century. Poems by monks of the time have been translated and show how impressive the displays must have been:

> *Equuleus is a little horse,*
> *Its shower is one to admire;*
> *There they go, some fast, some slow,*
> *Goodness gracious, great balls of fire!*

Unfortunately the shower is virtually extinct today, but if you're very lucky you may catch a glimpse of the odd green meteor around the peak of the display on 6 February.

LOCATION: Equuleus is happily eating sugar lumps on the right-hand side of the northern chart.

LATIN NAME
Equuleus
ENGLISH NAME
The Little Horse
ABBREVIATION
Equ
LATIN POSSESSIVE
Equulii

α STAR
Kitalpha
MAGNITUDE
3·9
STAR COLOUR
Yellow

An early 19th-century sky shown here around Equuleus contains four constellations that no longer exist – can you find them?

Andromeda

LATIN NAME
Andromeda

ENGLISH NAME
Andromeda

ABBREVIATION
And

LATIN POSSESSIVE
Andromedae

α STAR
Apheratz

MAGNITUDE
2.06

STAR COLOUR
White

Andromeda is the daughter of Cepheus who, thanks to the bragging of her mother, Cassiopeia, had to be chained to a rock as a sacrifice to the Sea Monster. Thankfully Perseus came along in the nick of time with the head of Medusa and turned the monster to stone. They married and lived happily ever after. Now back to reality...

LOCATION: Princess Andromeda branches off the top-left corner of Pegasus on the northern chart.

DEEP-SKY OBJECT
M31

TYPE
Galaxy

MAGNITUDE
4.8

SIZE
3°

DISTANCE IN LIGHT-YEARS
~2.8 million

The Great Nebula is the furthest object you can see reasonably clearly with your eye (see also M33 in Triangulum). Looking like an insignificant fuzzy patch, it is in fact a galaxy larger than our own, at a distance of over 2.8 million light-years (at current estimates). Of course you remember this distance is also 26 quintillion kilometres in Earth units! It's a mighty long way. Even so, it's enormous: 3° in length – that's six times the width of the Moon.

Here's a science-fiction view – it's only for fun. Andromeda is not this bright and the Moon can never be this close by (not unless there is some orbital catastrophe – maybe caused by a Martian invasion, nearby black hole or any other plausible scenario). The diagram is just here to convey the immense real size of the entire galaxy in the sky compared to the Moon.

Triangulum

An ancient constellation of three stars. And what else can you make with three stars? The Greeks knew it as Deltoton Ferdinandea, because it looked like the capital letter delta. Originally this was called Triangulum Major until Mr T. Minor (see below) left the scene.

Giuseppe Piazzi discovered the first asteroid in this constellation on 1 January 1801. Originally known as Ceres Ferdinandea, after the goddess and king of Sicily (where Piazzi's observatory was), its name was soon shortened to Ceres, as we know it today.

LOCATION: Triangulum is up on the left of the northern chart.

LATIN NAME
Triangulum
ENGLISH NAME
The Triangle
ABBREVIATION
Tri
LATIN
POSSESSIVE
Trianguli

α STAR
Rasalmothallah
MAGNITUDE
3.41
STAR COLOUR
White

DEEP-SKY OBJECT **M33**	OBJECT **ANTON 2**
TYPE Galaxy	TYPE Old constellation
MAGNITUDE 5.7	
SIZE 1°	
DISTANCE IN LIGHT-YEARS 3 million	

Witnesses claim to have seen this, the Pinwheel Galaxy, on extremely clear nights. I have not, and suspect those who say they have of drinking too much carrot juice. Having said that, if M33 is visible, it is the furthest object you can see with your eye – which, all told, is quite an impressive feature.

Here is the abandoned constellation of Triangulum Minor which is made of three fainter stars below the main constellation. Polish astronomer Johann Hevelius popped this faint group in the sky and it appeared on quite a few star atlases before disappearing into the backwaters. You see the problem three hundred years ago was that lots of famous astronomers were making star atlases, so the world became awash with different constellations. With only a limited amount of space, some less interesting groups fell by the wayside, and, let's face it, a small triangle is not that exciting (for Anton 3, see page 78.)

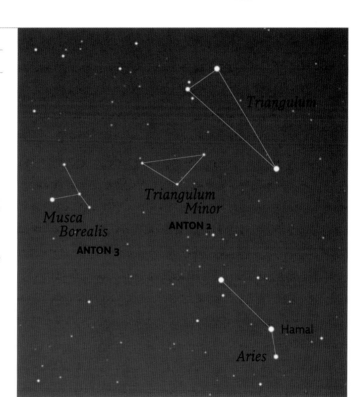

Aries

LATIN NAME
Aries

ENGLISH NAME
The Ram

ABBREVIATION
Ari

LATIN
POSSESSIVE
Arietis

α STAR
Hamal

MAGNITUDE
2.0

STAR COLOUR
Yellow-orange

Well, the designers were in full 'imagination' mode when they decided this constellation looked like a Ram. This Greek group depicts the animal with the Golden Fleece that Jason and his Argonauts were after.

The name Hamal is derived from the Arabic Al Ras al Hamal, meaning the Head of the Sheep.

LOCATION: Aries is munching on grass over on the far left of the northern chart.

OBJECT
ANTON 3

TYPE
Old constellation

Here is the abandoned constellation of Musca Borealis, the Northern Fly, which can be found on the chart with Anton 2 shown in Triangulum (see page 77). Eventually someone swatted the fly and it left our skies for good. The sky has a long history of groups that have come and gone – it really was just a case of later sky mappers not paying any attention to some of the earlier star groups.

Pisces

LATIN NAME
Pisces

ENGLISH NAME
The Fish

ABBREVIATION
Psc

LATIN
POSSESSIVE
Piscium

α STAR
Alrescha

MAGNITUDE
3.79

STAR COLOUR
White

An ancient Roman constellation, possibly that of Venus and her son Cupid. These two changed themselves into fishes so that they could swim away from the monster Typhon, (they couldn't stand the tea he made).

LOCATION: The Fish swim just left of the centre of the northern chart.

DOUBLE STAR
ρ and 94

MAGNITUDE
5.3 and 5.6

SEPARATION
7' 27"

COLOURS
Yellowish and golden

Aquarius

A very ancient group dating back to the Babylonians that has always been seen as a man pouring water from a jug. This could have had something to do with the rainy season when the group appeared at its best in the sky. All this part of the sky has a watery connection, under the control of Aquarius.

LOCATION: Aquarius is pouring water all over the bottom-right part of the northern chart.

LATIN NAME
Aquarius
ENGLISH NAME
The Water Bearer
ABBREVIATION
Aqr
LATIN POSSESSIVE
Aquarii

α STAR
Sadalmelik
MAGNITUDE
3.0
STAR COLOUR
Yellow

Cetus

An old group of the Perseus/Andromeda saga, for Cetus was the monster whom Poseidon had sent to nibble on Andromeda. Also known as the Sea Monster, this group makes up the fourth largest constellation and contains the first variable star of its class to be discovered – Mira.

LOCATION: Cetus is resting over on the lower left of the northern chart.

VARIABLE STAR
Mira
o Cet

MAGNITUDE RANGE
2.0 to 10.1

PERIOD
331.96 days

Apart from novae, Mira ('The Wonderful') was the first variable star to be identified – by Dutch astronomer David Fabricus in 1596. As a result, other long-period variables are also known as Mira-type stars. As Mira fluctuates so does its colour.

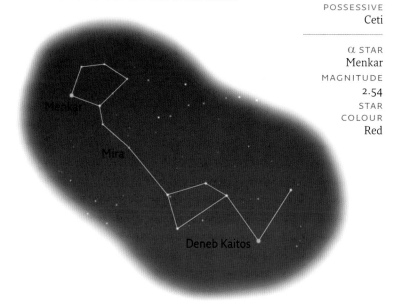

LATIN NAME
Cetus
ENGLISH NAME
The Whale
ABBREVIATION
Cet
LATIN POSSESSIVE
Ceti

α STAR
Menkar
MAGNITUDE
2.54
STAR COLOUR
Red

A couple of blobs and a black bit are just parts of an amazing vista, once you've escaped from the doom and gloom that is the South Celestial Pole.

The Southern Charts

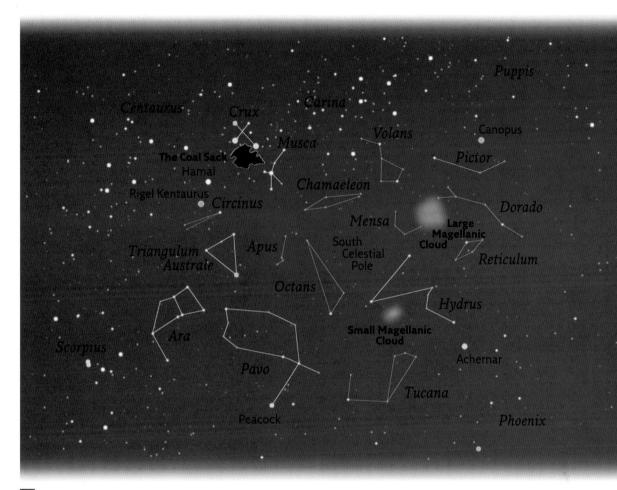

Galactic Cluster
Globular Cluster
Nebula
Galaxy

South Celestial Pole chart

It has to be said that the South Pole of the sky is less than exciting. I have already mentioned that no South Star exists, and we use the Crux and nearby friends to locate the Pole position. Not only is there no main star but there's not much at all around here – it's just very dark and mysterious. Having said that, a few hops away back towards Crux, we find the most magnificent skies anywhere. The southern night is truly awesome with its outstanding view of the Milky Way. Don't be taken in by my ramblings, though; if you can, go and have a look. The deep inky blackness, peppered with coloured stellar jewels, and the misty mysterious Magellanic Clouds. Oh yes, a sight to behold.

January to March Skies

From overhead (well, probably a little over to the north), the brilliant stars of Rigel, Sirius, Achernar and Canopus form the Great Southern Summer Curve (GSSC) – there, I've named an asterism! May it be remembered for ever more. To the left of the GSSC is the Milky Way – not that this is the best time of the year for milky viewing as we're looking out of our Galaxy towards empty space and not into it with all the stars, gas and dust that it comprises.

Those Magellanic Clouds are hanging just below my asterism. Orion is high, with Eridanus starting next to Rigel – this is a long flowing constellation that you can trace down to the bright star Achernar.

Star Sights

Large Magellanic Cloud
Sirius, the brightest star in the night sky
The Great Summer Southern Curve
α Centaurids meteor shower (peak around 8 February)
The Orion nebula, M42 (see page 45)

Galactic Cluster
Globular Cluster
Nebula
Galaxy

The southern
summer skies
looking south.

Canis Major

LATIN NAME
Canis Major
ENGLISH NAME
The Great Dog
ABBREVIATION
CMa
LATIN
POSSESSIVE
Canis Majoris

α STAR
Sirius
MAGNITUDE
-1.46
STAR COLOUR
White

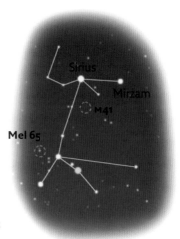

Here we find Sirius ('sparkling' or 'scorcher'), the brightest star in the sky after the Sun. Its fine sparkling appearance is due to its relative closeness – a mere 8.6 light-years away.

Looking closer we find that Sirius, also known as 'the Dog Star', is not alone in its corner of the Universe – this is a binary system. The companion star is quite small, only just over three times the diameter of the Earth. Its size and location lead to its being known as the 'Pup', although boring astronomers call it Sirius B. So, what of the Pup? Technically it is not an ordinary star but an object mysteriously called a white dwarf. These are the hot, dense, compact, glowing remains of stars like the Sun. Given enough time, a white dwarf will eventually

DEEP-SKY OBJECT **M41**	DEEP-SKY OBJECT **Mel 65**
TYPE Galactic cluster	TYPE Galactic cluster
MAGNITUDE 4.5	MAGNITUDE 4.1
SIZE 38"	SIZE 8'
DISTANCE IN LIGHT-YEARS 2,300	DISTANCE IN LIGHT-YEARS 5,000

In reality this is a happy family of around 100 stars of many colours.

Around 60 stars make up this, the τ Canis Majoris cluster, or call it NGC 2362 if you wish.

What a Great Dog!

cool down to become a dark black dwarf – a solid cold sphere roaming the cosmic wastelands until the ends of time. In the meantime Sirius and its 'Pup' are happily orbiting each other over a period of about 50 years.

The Egyptians knew Sirius as Sothis, the Nile Star, because when it appeared in the morning skies before the Sun, they knew that it was nearly time for the river's seasonal floods.

The Greeks did themselves proud with their Great Dog design, as it sort of, kind of, almost looks like a stick dog when you join up all the stars correctly.

LOCATION: The whole of the constellation is shown to the upper left on the southern chart.

Puppis

LATIN NAME
Puppis

ENGLISH NAME
The Stern

ABBREVIATION
Pup

LATIN POSSESSIVE
Puppis

ξ STAR
Naos

MAGNITUDE
Naos

STAR COLOUR
Bluish

This is one of the constellations that was ripped from the astral timbers of Argo Navis, the original ship of the Argonauts. The complete Argo Navis constellation was a fine sky-worthy vessel that sailed the stars through many a stormy night. Then, alas, along came South African astro-designer Nicholas La Caille, who did a dastardly chainsaw deed and created three groups we know today as Puppis, Carina and Vela. At the time there was no one in control of constellations, so if you wanted you could do pretty much anything you liked to the night sky – whether your designs were eventually accepted was another matter, but in this case Pirate Nick won the battle.

This is the most northerly part of the former 'ship' and, although it doesn't look much like a stern, the main stars are quite bright and easy to locate.

LOCATION: Puppis is on the middle left of the southern chart.

DEEP-SKY OBJECT
M47

TYPE
Galactic cluster

MAGNITUDE
4.4

SIZE
30'

DISTANCE IN LIGHT-YEARS
1,600

This group of around 50 stars looks like a fuzzy smudge, and then only if you have a really dark sky.

DEEP-SKY OBJECT
NGC 2451

TYPE
Galactic cluster

MAGNITUDE
2.8

SIZE
50'

DISTANCE IN LIGHT-YEARS
850

You can see this easily with the eye and yet apparently the great astronomers Charles Messier and William Herschel couldn't find it! Around 40 stars make up this group.

Dorado

Designed by those ever-friendly mariners Frederick de Houtman and Pieter Dirkszoon Keyser, Dorado is famous because it contains part of the Large Magellanic Cloud, a smaller satellite galaxy to our own.

History blurs where the 'Goldfish' title came from, and it must be said the Swordfish or indeed the Mahi-Mahi fish would be safer bets if we could travel back in time and ask Fred or Pete what sea creature gave them their inspiration.

LOCATION: Dorado is where the fuzzy patch is on the southern chart.

LATIN NAME	Dorado
ENGLISH NAME	The Goldfish
ABBREVIATION	Dor
LATIN POSSESSIVE	Doradus

α STAR	Bole
MAGNITUDE	3.3
STAR COLOUR	Bluish

DEEP-SKY OBJECT
NGC 2070
TYPE
Nebula
MAGNITUDE
5.0
SIZE
40' x 20'
DISTANCE IN LIGHT-YEARS
179,000

The Tarantula nebula is also called 30 Doradus as it was first catalogued as a star. This wondrous cloud must be classed in the top five of the nebulae charts – plus its real size is over 20 times that of the Orion nebula.

DEEP-SKY OBJECT
Large Magellanic Cloud
TYPE
Barred spiral galaxy
MAGNITUDE
0.4
SIZE
9°10' x 2° 50'
DISTANCE IN LIGHT-YEARS
179,000

One-quarter the size of our galaxy, this cloud, like the Small Magellanic, looks just like a large piece that has been taken away from the Milky Way and cast adrift. Known by those in the know simply as the LMC.

Reticulum

Isaak Habrecht of Strasbourg put this little group together in the 17th century. Originally it was simply a rhombus, but not satisfied with that, someone had to 'tinker' with it to show it as an instrument called a reticule – a device that astronomers stick in telescopes to help them measure star positions.

LOCATION: Reticulum is just above and to the right of the fuzzy patch on the southern chart.

LATIN NAME	Reticulum
ENGLISH NAME	The Net
ABBREVIATION	Ret
LATIN POSSESSIVE	Reticuli

α STAR	α Ret
MAGNITUDE	3.4
STAR COLOUR	Yellow

DOUBLE STAR
ζ Ret
MAGNITUDE
5.2 & 5.5
SEPARATION
5' 10"
COLOURS
Both yellow

April to June Skies

Depending on how far south you are (the further south the better), the Milky Way at this time of year can really span the sky high overhead. This majestic band is splattered by some excellent bright stars of the constellations Centaurus, Crux, Carina, Vela and Canis Major. Meanwhile the Large and Small Magellanic Clouds smudge the sky in the far south. If you are in 'imagination' mode, why not try to join up the four constellations of Carina (the Keel), Vela (the Sails), Puppis (the Stern) and Pyxis (the Compass) into the original vast magnificent ship, Argo Navis. Meanwhile, over to the north is the sea snake, Hydra. This isn't particularly bright, but its scaly length covers an incredible amount of the sky.

Star Sights

The Coal Sack dark nebula
The Jewel Box cluster, NGC 4755
The Eta Carina nebula, NGC 3372
The globular cluster omega Centauri, NGC 5139
η Aquarids meteor shower

 Galactic Cluster
Globular Cluster
Nebula
Galaxy

The southern
autumn skies looking
south.

Hydra

LATIN NAME
Hydra
ENGLISH NAME
The Water Snake
ABBREVIATION
Hya
LATIN POSSESSIVE
Hydrae

α STAR
Alphard
MAGNITUDE
2.0
STAR COLOUR
Orange

A nasty nine-headed (!) monster who eventually came across Hercules and met a sticky end. The largest constellation, Hydra is extremely long and borders 14 other constellations – that's more than any other. The name of Alphard, its leading star, means 'heart of the snake'.

LOCATION: The top of Hydra is beneath Leo on our northern chart (see page 53), and the rest of it is on the southern.

Spica

Alphard

R Hya

VARIABLE STAR
R Hya
MAGNITUDE RANGE
4.5 to 9.5
PERIOD
389 days

This is a Mira-type (long-period) variable.

No wonder Hydra wasn't a pleasant beast; look at what he had to cope with: a large cup, a sextant, a crow and an owl all sitting on his back while all he wanted to do was a few quiet lengths of the local stellar pool. Actually, Noctua, the Owl, is no longer a constellation, nor is Felis, the Cat.

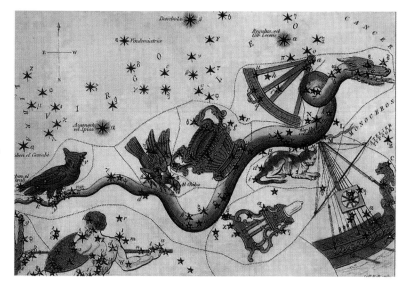

Centaurus

Not many who met Hercules, or even those in the same vicinity, had a good time. Centaurus was a case in point: this was Chiron, who was accidentally killed by a mis-shot arrow from our hero's bow. In mythology, centaurs were considered smelly and not the most pleasant of company, but Chiron deserves a better press – he was fun-loving and very scholarly, teaching many Greek heroes.

Very close to Rigel Kentaurus ('the foot of the Centaur') is the very small and faint Proxima Centauri: the nearest star to us after the Sun. It is a mere 4.25 light-years away, which in a useless measurement is 40,018,760,000,000 kilometres. There are those who think Proxima is the tiniest distant member of a Rigel Kentaurus system of three stars. And by distant I mean Proxima may be the equivalent of over 250 times Pluto's distance from our Sun.

LOCATION: The mighty Centaur teaches those who will listen just to the lower left of the middle of the southern chart.

LATIN NAME
Centaurus
ENGLISH NAME
The Centaur
ABBREVIATION
Cen
LATIN POSSESSIVE
Centauri

α STAR
Rigel Kentaurus
MAGNITUDE
−0.01
STAR COLOUR
Yellow

DEEP-SKY OBJECT
NGC 5139
TYPE
Globular cluster
MAGNITUDE
3.65
SIZE
36'
DISTANCE IN LIGHT-YEARS
17,000

This is omega Centauri – 'That's a star!' I hear you cry. How can a star be the best globular cluster in the sky? Well, it's because the true nature of this mysterious object wasn't known until the invention of the telescope – it simply looks like a star. A bit of a fuzzy star, I admit.

DEEP-SKY OBJECT
NGC 3766
TYPE
Galactic cluster
MAGNITUDE
5.3
SIZE
12'
DISTANCE IN LIGHT-YEARS
5,500

Mr La Caille had a browse from South Africa back in the early 1750s and found this cluster. It's been called 'attractive' in its time, and is quite colourful when seen through binoculars.

Crux

LATIN NAME
Crux
ENGLISH NAME
The Cross
ABBREVIATION
Cru
LATIN POSSESSIVE
Crucis

α STAR
Acrux
MAGNITUDE
0.9
STAR COLOUR
Bluish

'Oh, those four stars will do,' said Johann Bayer, taking a few stars from neighbouring Centaurus to make the Southern Cross. He then built this smallest constellation into his star book *Uranometria* (1603) and it became a 'fixed' constellation from then on.

Like the Plough in the north, the Crux's easily identifiable pattern made it familiar to many cultures: some Aboriginal tales describe it as two cockatoos sitting in a gum tree, while in southern Africa it could be linked to the two bright stars of Centaurus next door to make a giraffe.

If I were going to score all the constellations for majestic 'wowness', then Crux would have to come pretty high. There are so many things going on, what with

The busy area around Crux, with the bright Milky Way running through the view. (Image courtesy of Peter Michaud)

DEEP-SKY OBJECT
NGC 4755

TYPE
Galactic cluster

MAGNITUDE
4.2

SIZE
10'

DISTANCE IN
LIGHT-YEARS
7,600

So useful was the Southern Cross to the mariners of ye olden dayes that this recognisable group found its way on to the flags of Australia, New Zealand, Papua New Guinea and Samoa.

The Jewel Box cluster is a 'gem' of a group. Its stars glisten like a box of multi-coloured jewels, with blues, reds and whites galore. This cluster, sitting around κ Cru, is right next door to the Coal Sack dark nebula.

the Milky Way, Coal Sack, Jewel Box, dust, gas, stars, starry clusters, more dust, more stars, more gas, more clusters, etc. In fact, for shiny super-greatness my top five constellations probably would be, in no particular order: Crux, Centaurus, Carina, Sagittarius and Scorpius – all of which need a good southerly location if you are to see them well.

LOCATION: The Southern Cross is nestled underneath Centaurus, next to the dark patch, on our southern chart.

DEEP-SKY OBJECT
The Coal Sack

TYPE
Dark nebula

MAGNITUDE
6°

SIZE
30" x 5°

DISTANCE IN
LIGHT-YEARS
550

This 'hole' in the flowing Milky Way is due to a cloud of dust and gas that is blocking out the star-light from behind. The Coal Sack is probably the closest dark nebula to us.

Here's the central part of the picture opposite with locators for some wonderful deep-sky objects. The two bright stars of alpha and beta Centauri are off to the left. (Image courtesy of Peter Michaud)

Vela

LATIN NAME
Vela

ENGLISH NAME
The Sails

ABBREVIATION
Vel

LATIN POSSESSIVE
Velorum

α STAR
γ Vel

MAGNITUDE
1.8

STAR COLOUR
Bluish

If you've read about Puppis (see page 86), you know the story. Argo Navis was a fine vessel. In mythology Jason sailed in her with his Argonauts hither and thither. Over the seven seas the creaking timbers survived well until the star-mapper Nic 'Pirate' La Caille carved her up into three pieces during the 'Battle of the Stars'. This was during that time in the 1700s and 1800s when the only thing astro-designers wanted was for one of their groups to be accepted into sky lore. Nic was fortunate this time, while many around him 'walked the plank' into the murky waters of 'constellation history'.

As far as Argo Navis is concerned, it has been said that on dark, calm nights, you can still hear the creakin' of the yardarm from that fateful break-up. Because of this heavenly 'shipwreck', Vela has no alpha or beta stars, but is led by the unnamed gamma Velorum – sounds like a magic spell if you say it out loud.

Interesting Milky Way fact: our flowing friend travels through Vela. Nothing of note there, you may think. However, this is the only part of its entire circuit around the sky that is broken. Across the Milky Way here runs a ribbon of dark dust and gas that cuts our river of light completely.

LOCATION: The sails are flapping in the wind just down from centre-right of the southern chart.

DEEP-SKY OBJECT
IC 2391

TYPE
Galactic cluster

MAGNITUDE
2.5

SIZE
50'

DISTANCE IN LIGHT-YEARS
580

This family of about 30 stars is centred around o Vel.

DEEP-SKY OBJECT
NGC 2547

TYPE
Galactic cluster

MAGNITUDE
4.7

SIZE
20'

DISTANCE IN LIGHT-YEARS
1,950

A fine group of around 50 stars discovered by La Caille. Yes, he who did the 'carving up the ship' deed.

The three pieces of Argo Navis are now: Vela (The Sail), Puppis (The Stern) and Carina (The Keel). Pyxis (The Compass) is commonly included, but was not a part of the original ship. Here is the full-blown vessel as seen in Hevelius's *Uranographia*.

Carina

Part of the old big constellation that was Argo Navis. See Vela for the full story of what happened on one 'dark and stormy night' when this good ship sailed into the path of pirate Nic 'Black Beard' La Caille and became three new constellations.

Canopus is the second brightest star of the night sky, being beaten by Sirius, in Canis Major. The origin of its name, like many that are millennia old, is bathed in murky mystery. It could possibly have come from the Egyptian words for 'golden earth' – *karl nub* – due to its pale yellow colour. I say yellow with caution, as it seems some books consider it whitish-blue! Take a look and decide for yourself.

LOCATION: The Keel sits adrift down on the lower right of the southern chart.

LATIN NAME
Carina
ENGLISH NAME
The Keel
ABBREVIATION
Car
LATIN POSSESSIVE
Carinae

α STAR
Canopus
MAGNITUDE
–0.72
STAR COLOUR
Yellowish

Beware: Crux, the smallest constellation, has four bright stars that sit close to the two very bright leading stars in Centaurus. However, just along the way are four stars arranged in a vaguely similar pattern – hence their nickname of the False Cross. Some people have been known to come unstuck at this point. Just remember that this group is not as bright as the true Southern Cross and does not have two shiny stars next door.

DEEP-SKY OBJECT **Mel 82**	DEEP-SKY OBJECT **NGC 3114**	DEEP-SKY OBJECT **IC 2581**	DEEP-SKY OBJECT **IC 2602**	DEEP-SKY OBJECT **Mel 103**	DEEP-SKY OBJECT **NGC 3372**
TYPE Galactic cluster	TYPE Galactic cluster	TYPE Galactic cluster	TYPE Galactic cluster	TYPE Galactic cluster	TYPE Nebula
MAGNITUDE 3.8	MAGNITUDE 4.2	MAGNITUDE 4.3	MAGNITUDE 1.9	MAGNITUDE 3.0	MAGNITUDE 5.0
SIZE 30'	SIZE 35'	SIZE 8'	SIZE 50'	SIZE 55'	SIZE 2°'
DISTANCE IN LIGHT-YEARS 1,300	DISTANCE IN LIGHT-YEARS 3,000	DISTANCE IN LIGHT-YEARS 2,868	DISTANCE IN LIGHT-YEARS 479	DISTANCE IN LIGHT-YEARS 1,300	DISTANCE IN LIGHT-YEARS 10,000

Also designated as NGC 2516, this is a bright cluster of about 100 stars.

Apparently there are 171 stars in this cluster. How anyone can be that precise is a mystery.

A cluster of about 25 stars.

I could have used the Melotte designation here of Mel 102, for this fine cluster that has affectionately been called the Southern Pleiades.

Also catalogued as NGC 3532, this starry group is near the Eta Carina nebula (NGC 3372). It sits in a very busy part of the Milky Way, so add binoculars and the scene is marvellous.

The Eta Carina nebula is one of the great star-forming regions in our galaxy. Here one of the most massive stars we have ever detected was created – Eta Carina; hence the name.

SOUTHERN July to September Skies

It's 'full glory of the Milky Way' time as it flows north to south overhead. There are so many objects to view it's difficult to find a place to start! I guess the Sagittarian teapot and the tail of Scorpius lead the field with clusters galore, as well as the general mix of distant milky stars meandering within bright and dark clouds of dust and gas. A truly dark sky is needed to appreciate the full magnificence of this sight. Not surprisingly there are emptier skies to either side – only Fomalhaut and Achernar light up the southeast, while Spica is setting over in the west.

Star Sights

The Milky Way!

M24, star clouds in Sagittarius

Table of Scorpius cluster, NGC 6231

Zubenelgenubi, the leading star of Libra

*Arkab Prior and Arkab Posterior, double star in
 Sagittarius*

*δ Aquarids meteor shower (two peaks around 29 July
 and 8 August)*

⬡ Galactic Cluster
⊕ Globular Cluster
▢ Nebula
⬭ Galaxy

The southern winter
skies looking south.

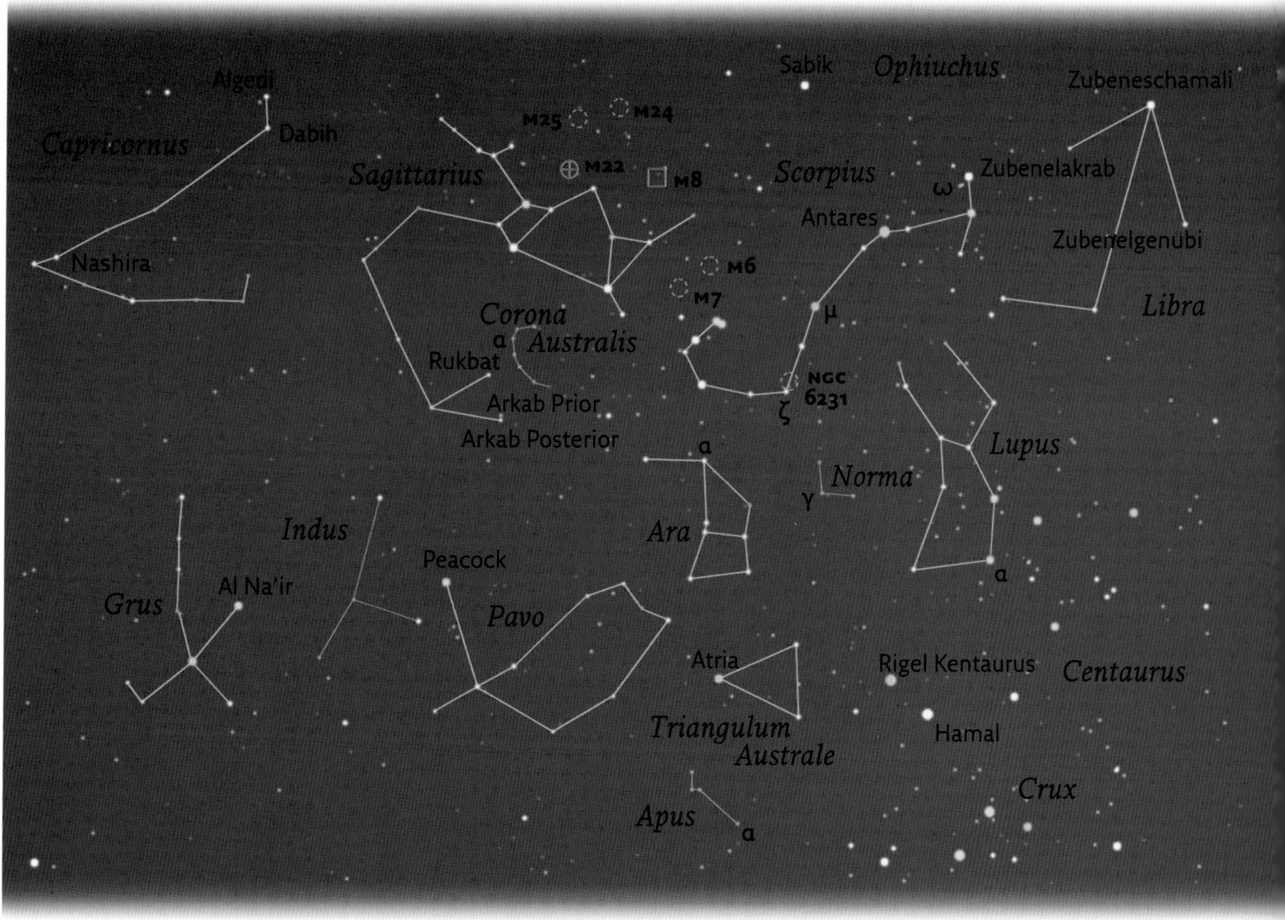

Capricornus

LATIN NAME
Capricornus
ENGLISH NAME
The Sea Goat
ABBREVIATION
Cap
LATIN
POSSESSIVE
Capricornus

α STAR
Algedi
MAGNITUDE
3.6
STAR COLOUR
Yellow

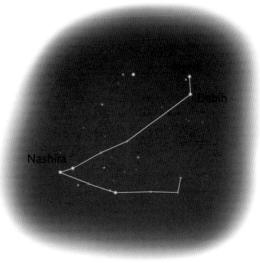

A very old constellation, maybe with Oriental roots, of the half-goat half-fish – now there's a combination of animals. According to reliable Roman sources, this was the god Pan, who became a bit fishy after diving into the Nile to escape from the Typhon tea monster. But apparently only the bit that got wet was changed into a fish. Well, that's much clearer.

Look around Capricorn in the sky and you'll find it's all pretty watery – there are Aquarius, Pisces, Cetus and Piscis Austrinus. There is an ancient link in as much as this time of year was associated with rains and floods.

AMAZING FACT: Capricornus is the smallest of the 12 signs of the zodiac.

LOCATION: Our watery champ swims on the top left of the southern chart.

DOUBLE STAR
Algedi
α Cap
MAGNITUDE
4.2 & 3.6
SEPARATION
6' 18"
COLOURS
Both goldishy (as opposed to goldfishy)

I'm not keen on those goaty-fish eyes that keep staring at me.

Libra

Libra was not there before Roman times – the claws of Scorpius, the Scorpion, were. So, what happened? Well, without a care in the world, the Romans cut off the claws and made a nice pair of Scales. The simple life indeed. There was no comeuppance – by the time anyone noticed, 1,500 years had gone past.

Don't get too excited; this is not a startling constellation, but it's worth mentioning simply because of the marvellous names of its stars: Zubenelgenubi, Zubeneschamali, Zubenelakrab and Zubenelakribi.

Zubeneschamali (β) is apparently the greenest star you can see with the eye.

LOCATION: The Scales are on the top right of the southern chart.

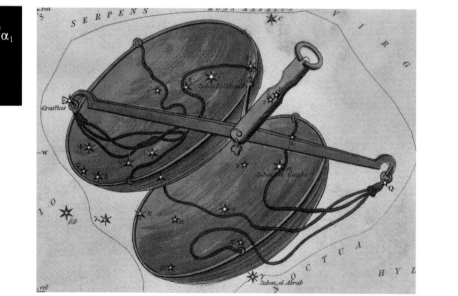

LATIN NAME
Libra
ENGLISH NAME
The Scales
ABBREVIATION
Lib
LATIN POSSESSIVE
Librae

α STAR
Zubenelgenubi
MAGNITUDE
2.75
STAR COLOUR
White

DOUBLE STAR
Zubenelgenubi
α Lib and 8 Lib
MAGNITUDE
2.8 & 5.2
SEPARATION
3' 51"
COLOURS
Bluish and white

VARIABLE STAR
Zubenelakribi
δ Lib
MAGNITUDE RANGE
4.9 to 5.9
PERIOD
2.327 days

Libra, the Scales. The name of the leading star Zubenelgenubi means 'the Southern Claw of the Scorpion', and indicates the ancient past when Libra was a part of Scorpius.

Scorpius

LATIN NAME
Scorpius
ENGLISH NAME
The Scorpion
ABBREVIATION
Sco
LATIN
POSSESSIVE
Scorpii

α STAR
Antares
MAGNITUDE
0.96
STAR COLOUR
Red

Look out Orion, Apollo has sent this stinging nasty to do you in! This is why Orion and Scorpius are placed on opposite sides of the sky, so the Hunter doesn't have to put up with any more trouble.

Although from mid-northern latitudes you do get to see the brilliant Antares ('the rival of Mars'), the splendour of Scorpius is lost unless you travel further south. The whole curvy sweeping S-shape is visible from latitudes less than 40°N such as Madrid, Spain, Naples, Italy; New York and Salt Lake City, USA; Ankara, Turkey; and Beijing, China.

The Scorpion used to have wonderful claws in the sky until they were cut off by the Romans to make a 'new' constellation of Libra, the Scales (see page 99).

LOCATION: The Scorpion is to the upper-middle-right of the southern chart.

DEEP-SKY OBJECT
M7
TYPE
Galactic cluster
MAGNITUDE
3.3
SIZE
1° 20'
DISTANCE IN
LIGHT-YEARS
800

Sometimes known as Ptolemy's cluster after he who described it back in AD 130. This is probably the finest deep-sky object in the constellation, and look at its size – more than twice as big as the Moon!

DEEP-SKY OBJECT
M6
TYPE
Galactic cluster
MAGNITUDE
4.2
SIZE
20'
DISTANCE IN
LIGHT-YEARS
2,000

The Butterfly cluster is a fine group of around 80 stars.

DOUBLE STAR
μ Sco
MAGNITUDE
3.0 & 3.6
SEPARATION
5' 30"
COLOURS
Both bluish

DOUBLE STAR
ζ1 and δ2
Sco
MAGNITUDE
3.6 & 4.9
SEPARATION
6' 30"
COLOURS
Orange and bluish

DEEP-SKY OBJECT
NGC 6231
TYPE
Galactic cluster
MAGNITUDE
2.5
SIZE
15'
DISTANCE IN
LIGHT-YEARS
5,900

These two form the lower part of the Table of Scorpius. It's all a great part of the sky.

DOUBLE STAR
ω1 and ω2 Sco
MAGNITUDE
4.1 & 4.6
SEPARATION
14' 30"
COLOURS
Blue and orange

Sagittarius

He may well have a bow, but when does an archer look like a teapot? When he is Sagittarius. Just join up the stars and you definitely get something for making tea in.

This is a constellation that seems to be a mix of civilisations. There could be Sumerian and Greek influences, while the name is Roman. Gazing into this part of the sky, you are looking directly into the heart of our Galaxy. This means the whole area is the best bit of the Milky Way, full of nebulae, star clusters and dust clouds, as you can see from the list and images overleaf.

LOCATION: To the top of the middle on the southern chart is where you'll find Sagi with his bows and arrows.

LATIN NAME
Sagittarius
ENGLISH NAME
The Archer
ABBREVIATION
Sgr
LATIN POSSESSIVE
Sagitarii

α STAR
Rukbat
MAGNITUDE
3.97
STAR COLOUR
Bluish-white

Rukbat
Arkab Prior
Arkab Posterior

DEEP-SKY OBJECT
M8

TYPE
Nebula

MAGNITUDE
5.8

SIZE
1° 30' x 40'

DISTANCE IN LIGHT-YEARS
5,200

The Lagoon nebula was noted by Le Gentil in 1747. Faint, yes, but just visible to your eye on the clearest of crispy nights. It gets its name from a dark, lagoonish band of dust that meanders across the cloud, visible via telescopes.

DEEP-SKY OBJECT
M22

TYPE
Globular cluster

MAGNITUDE
5.1

SIZE
24'

DISTANCE IN LIGHT-YEARS
10,000

This object may have been first noted by Mr Abraham Ihle as long ago as 1665. In fact M22 could have been the very first globular cluster identified. It ranks alongside M13 in impressiveness (see page 68).

DEEP-SKY OBJECT
M24

TYPE
Star clouds

MAGNITUDE
4.5

SIZE
1° 30'

DISTANCE IN LIGHT-YEARS
10,000

The Sagittarius star cloud is a patch of fuzzy light that is just a little brighter than the general mish-mash that is the Milky Way band. At three times the diameter of the Moon, it's quite a whopping good size too.

DOUBLE STAR
Arkab Prior and Arkab Posterior
β1 and β2 Sgr

MAGNITUDE
4.0 & 4.3

SEPARATION
28.3'

COLOURS
Bluish and white

No apologies – there is no diagram for these stars purely because they are too far apart to fit on the page! Being such an extremely wide double, they are clearly identified on the main chart.

In very dark clear skies this central area of the Milky Way is an impressive sight indeed. (Image courtesy Konrad Malin-Smith)

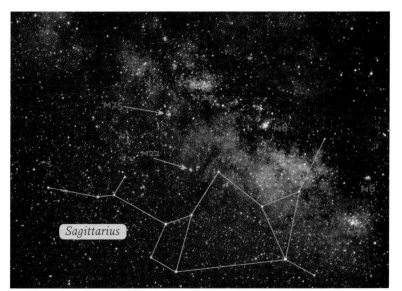

And with not too much difficulty the starry sights reveal themselves.

M25

M22

Sagittarius

M6

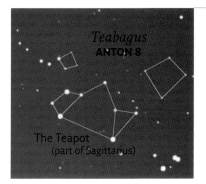

Teabagus
ANTON 8

The Teapot
(part of Sagittarius)

OBJECT
ANTON 8

TYPE
Old constellation

Hovering above the teapot that is Sagittarius is the abandoned constellation of Teabagus. (Yes, really. In the 18th century, due to the classical teapot shape of Sagittarius, one extrovert designer attempted to introduce the teabag!)

Corona Australis

This is a Greek design depicting the crown worn by Sagittarius, the constellation which sits next door. Therefore an earlier alternative Roman name for this curvy southerly group was Corona Sagittarii. The Milky Way, running through the area, makes it all the more interesting.

LOCATION: Look for the Southern Crown underneath the 'teapot' of Sagittarius.

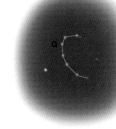

LATIN NAME
Corona Australis
ENGLISH NAME
The Southern
Crown
ABBREVIATION
CrA
LATIN
POSSESSIVE
Coronae Australis

α STAR
α CrA
MAGNITUDE
4.1
STAR COLOUR
White

Triangulum Australe

Three stars: what shall we make? The 16th-century Dutch mariners Frederick de Houtman and Pieter Dirkszoon Keyser had no problem deciding on a…triangle! However, this could be a much older group due to its easily identifiable stars – much more so than the northern hemisphere's equivalent, Triangulum.

LOCATION: The Triangle weaves its geometric magic in the lower-middle-right part of the southern chart.

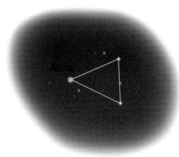

LATIN NAME
Triangulum
Australe
ENGLISH NAME
The Southern
Triangle
ABBREVIATION
TrA
LATIN
POSSESSIVE
Trianguli
Australis

α STAR
Atria
MAGNITUDE
1.9
STAR COLOUR
Orange

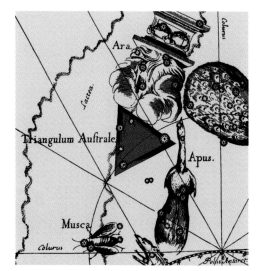

Triangulum Australe next to Apus and Ara as shown in the Hevelius star atlas *Uranographia*.

October to December Skies

We lose the brightest part of the Milky Way at this time as the Earth travels around the Sun carrying Sagittarius and Scorpius down towards the west. High in the middle-ish of the darkness is the bright star Fomalhaut, with the cheeky Achernar down and to the left a bit (depending on where and when of course). Looking east, a few cheeky bright chappies are rising: Canopus, and later Sirius. Apart from that it's fairly quiet – oh, there are the magnificent Magellanic Clouds, at their highest as we get into December. By the way, have you noticed how many bird constellations are flapping around at this time of year?

Star Sights

The Small Magellanic Cloud
Tracing the entire path of Eridanus
Mira, the famous variable star in Cetus
Orionids meteor shower (peak around 21 October)
The Large Magellanic Cloud (see page 83)

⊡ Galactic Cluster
⊕ Globular Cluster
☐ Nebula
⬭ Galaxy

The southern spring
skies looking south.

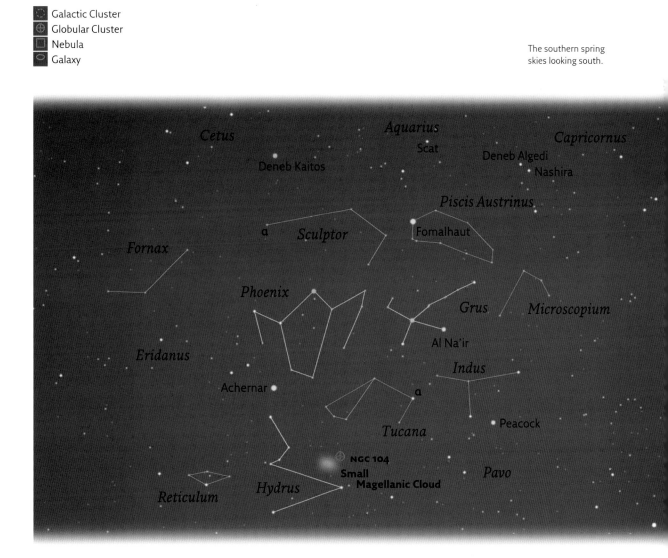

Eridanus

LATIN NAME
Eridanus
ENGLISH NAME
The River
ABBREVIATION
Eri
LATIN POSSESSIVE
Eridani

α STAR
Achernar
MAGNITUDE
0.5
STAR COLOUR
Bluish

Flowing over a very large area of the sky, this is an ancient constellation that could be the river created by Phaeton, the son of the Sun God, while links with the Euphrates and Nile are only to be expected. Meander all the way down and you'll find Achernar, an Arabic name meaning 'end of the river'. Once you've reached it (if you're far enough south) then you are gazing at the ninth brightest star in the night sky.

Epsilon (ε) is the third closest star to us (after Rigel Kentaurus and Sirius), at just 10.7 light-years. This star may also have planets with aliens on them.

LOCATION: Eridanus starts immediately to the right of the bright star Rigel in Orion on the southern January to March sky chart.

A sky chart from the 1820s showing the 'lost' constellation of Psalterium Georgii, George's Lute.

OBJECT
ANTON 6
TYPE
Old constellation

The star 53 Eri is known as Sceptrum after the wonderfully named abandoned constellation of Sceptrum Brandenburgicum, the Sceptre of Brandenburg.

OBJECT
ANTON 7
TYPE
Old constellation

Up to the right a bit (down to the left if you live in the southern hemisphere) is another forgotten group, Psalterium Georgii or George's Lute. This was a design by Abbé Maximillian Hell in

1789 to honour King George II. According to legend, the reason it disappeared as a constellation was partly due to an accident. The lute was carelessly left on the banks of Eridanus, where it was washed away during a heavy storm into the waters of time. A lesson for us all, I feel.

Octans

Nicholas La Caille constructed this constellation from the unbelievably faint stars around the South Pole of the sky back around 1751. Why he designed an octant, which is an instrument used for navigation, when there is no possible way this constellation could ever be used by mariners to find their way will forever remain a secret only known to mad Nic himself. At present, the star that finds itself the closest to the Pole is Sigma Octantis, which at magnitude 5.45 does not qualify for the title Polaris Australis.

LOCATION: Octans is at the centre – not surprisingly really – of the South Celestial Pole chart (see page 81).

LATIN NAME
Octans
ENGLISH NAME
The Octant
ABBREVIATION
Oct
LATIN POSSESSIVE
Octantis

α STAR
α Oct
MAGNITUDE
5.15
STAR COLOUR
Yellow

Tucana

Frederick de Houtman and Pieter Dirkszoon Keyser designed this bird while St Elmo's Fire darted about the rigging one clear sparkling night somewhere in the southern oceans. Oh, those were easy days. Their friend Johann Bayer placed their group in his Uranometria star atlas.

LOCATION: Look for the fuzzy blob at the bottom of the southern chart – that's where the Toucan is.

LATIN NAME
Tucana
ENGLISH NAME
The Toucan
ABBREVIATION
Tuc
LATIN POSSESSIVE
Tucanae

α STAR
α Tuc
MAGNITUDE
2.9
STAR COLOUR
Orange

DEEP-SKY OBJECT
Small Magellanic Cloud
TYPE
Irregular galaxy
MAGNITUDE
2.3
SIZE
5°19' x 3°25'
DISTANCE IN LIGHT-YEARS
196,000

Also catalogued as NGC 292, the SMC is one sixth the size of our Galaxy. This local little galaxy has been known through the ages, but particularly came to notice when Magellan sailed round the world in 1519. You could describe it as looking like a bit of the Milky Way that's broken off and is now roaming the starry skies in search of friendship. Just next door is the object mentioned right: NGC 104.

DEEP-SKY OBJECT
NGC 104
TYPE
Globular cluster
MAGNITUDE
4.0
SIZE
30'
DISTANCE IN LIGHT-YEARS
13,400

Crazy old astronomers thought this was a star –47 Tucanae to be precise. Then came modern-fangled telescopes and it became a big fuzzy blob!

The Toucan and the 'Nebuc Minor' (now the Small Magellanic Cloud) as seen on the *Uranographia* star atlas of Johann Hevelius.

Sun, Moon
and Planets

The Moon

Just glancing up at the Moon you cannot fail to see that it's made up of light and dark patches. Those olden astronomers took the darker areas to be seas against the brighter land. Even though we now know that this is not the case, the names of the 'seas' and watery-like features have remained in use, as you can see here.

The Moon dates from the formation of the Earth about 4.6 billion years ago. The most popular theory of how it was made involves a large object smashing into the Earth and blasting some of our planet, mixed with this object, into space, all of which rocky material formed a ring around us. Over a relatively short time, maybe just one year, this material clumped together to form the Moon. Personally, I think the Moon People made it out of green cheese (the Lunar-Cheddar Theory), but apparently I'm in the minority.

Have you ever wondered why the Moon has so many craters and yet the Earth has so few? We have to look to the early solar system for the answer. During this far-off time lots of objects were flying here, there and everywhere, smashing into everything that got in the way. The Earth received its fair share of knocks, but due to our weather, water and continental drift, virtually all of our early craters have since been wiped clean off the surface. Not so on the Moon; it has no atmosphere – it's too small to hold on to one of any significance. So every Moon-thing was just left in perfect condition, craters and all.

Sinus Aestuum	Bay of Heats
Mare Anguis	Serpent Sea
Mare Australe	Southern Sea
Mare Cognitum	Sea of Thoughts
Mare Crisium	Sea of Crisis
Palus Epidemiarum	Marsh of Epidemics
Mare Foecunditatis	Sea of Fertility
Mare Frigoris	Sea of Cold
Mare Humboldtianum	Humboldt's Sea
Mare Humorum	Sea of Humours
Mare Imbrium	Sea of Showers
Mare Insularum	Sea of Isles
Sinus Iridum	Bay of Rainbows
Mare Marginis	Marginal Sea
Sinus Medii	Central Bay
Lacus Mortis	Lake of Death
Mare Moscoviense	Moscow Sea
Palus Nebularum	Marsh of Mists
Mare Nectaris	Sea of Nectar
Mare Nubium	Sea of Clouds
Mare Orientale	Eastern Sea
Oceanus Procellarum	Ocean of Storms
Palus Putredinis	Marsh of Decay
Sinus Roris	Bay of Dews
Mare Serenitatis	Sea of Serenity
Mare Smythii	Smyth's Sea
Palus Somnii	Marsh of Sleep
Lacus Somniorum	Lake of the Dreamers
Mare Spumans	Sea of Foam
Mare Tranquilitatis	Sea of Tranquillity
Mare Undarum	Sea of Waves
Mare Vaporum	Sea of Vapours

Mare Frigoris

Plato

Mare Imbrium

Aristillus

Eratosthenes

Aristarchus

Archimedes

Mare Serenitatis

Oceanus Procellarum

Mare Crisium

Kepler

Mare Vaporum

Mare Tranquilitatis

Copernicus

Pallas

Grimaldi

Mare Nubium

Mare Foecunditatis

Ptolemeus

Hipparchus

Albategnius

Alphonsus

Mare Nectaris

Arzachel

Purbach

Tycho

Anton's official map of the Moon.

Clavius

The lunar seas

The *maria* ('seas') on the Moon were formed during the heaviest cratering sessions when the surface was cracked open, causing the molten material from below to leak out to form these vast, dark, lava-lake landscapes. The seas are where to land if you're going to the Moon, as these are the smoothest places. Indeed, you'll find that most of the Apollo missions of the late '60s and '70s landed there.

The *craters* are mainly caused by impacts from comets and asteroids that smashed into the Moon, temporarily melting the surface. Just like throwing a stone into a pond, where a ripple of water is made, the Moon rock splashes outwards too, but unlike water this rock solidifies quickly and we are left with a crater – in essence, a frozen ripple.

Can you see the combination of light and dark areas that make the 'Man in the Moon'?

Now look at this great picture of the waxing gibbous Moon. (Image courtesy Paul Whitmarsh)

The motion of the Moon

The Moon orbits the Earth in exactly the same time as it takes to rotate once on its axis – this is known as *synchronous rotation*. It's not an uncommon phenomenon, either: Titan around Saturn, Triton around Neptune and Io, Europa, Ganymede and Callisto around Jupiter are only a few examples of satellites that do the same thing. Synchronous rotation means that we always see the same side of the Moon – simply known as the 'near side'.

It's amazing to think we knew nothing about what the 'far side' looked like until the Russian space probe Luna 3 took the first pictures of it back in 1959. Those images showed that it too is covered in craters, but there are no real 'seas'.

Anyway, what I have just told you is not quite accurate – because there is an amount of wobbling around known as **libration** which actually allows us to see a little more than just one half of the Moon.

Another Moon word is *lunation*, which is the time it takes for the phases to repeat themselves, from one Full Moon to the next, or from one New Moon to the next, for example. This time is 29.5 days (called the **synodic month**). So if you see a Full Moon, the next Full Moon will be 29.5 days later. It is no coincidence that the synodic month is about the same length as one calendar month. Indeed this is where the word *month* comes from – originally it would have been *mooneth*.

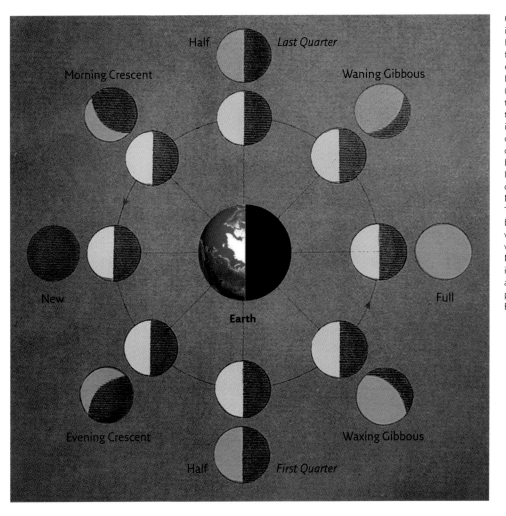

Half   Last Quarter

Morning Crescent   Waning Gibbous

New   Earth   Full

Evening Crescent   Waxing Gibbous

Half   First Quarter

Over 29.5 days there is a complete cycle of Moon phases as our friendly lunar companion orbits the Earth anticlockwise (as seen from above the North Pole). In this diagram the Sun is off to the left, constantly lighting up one half of the Moon, but how much of this lighted half we see is down to where the Moon is in its orbit. The outer, light-brownish Moons, with labels, show what phase the Moon is at when the inner white Moons are at various positions around the Earth.

Observing the Moon

Learning about the Moon involves the same process as learning the constellations: if you take it slowly, you'll soon be able to find your way around with ease.

The map on page 110 will help you to identify the light and dark areas of the Moon that you can see simply by gazing up into the sky. However, as with all observing, look for a little longer and more details will be revealed to

your starry (or moony) eyes: bright rays caused by ancient impacts have splattered material across the lunar globe. Then there are craters: three known as Copernicus, Aristarchus and Kepler stand out particularly well as they all sit in the centre of bright splodges in the dark Ocean of Storms (or Oceanus Procellarum).

The time to look for the features just mentioned is around Full Moon, but the crescent, half and gibbous phases also provide interesting viewing. (Gibbous is the phase between half moon and full moon.) Pay particular attention to the boundary between the sunlit and the dark part known menacingly (so this needs to be spoken in a deep, slow, B-movie kind of way) as the *terminator*.

The terror having now subsided, this boundary is where we can sometimes witness the Sun just lighting the features on either side. This allows the eye to catch tantalising hints of the mountains, craters, ridges and valleys. At times this terminator can demonstrate how bumpy the surface of the Moon is by the jagged appearance along its length. And there's more: because the Moon is constantly moving around the Earth, its phases are constantly changing. In the same way, therefore, the terminator is constantly displaying different highlights and shadows. Take a look if you don't believe me.

The place where dark meets light is a very interesting one to watch, as it can reveal the surface features of the Moon as appearing quite 'jagged' at times.

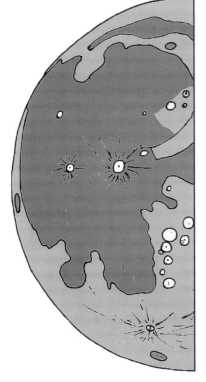

The Half Moon on the right is the one you see in the afternoon a week before Full Moon, whereas the one on the left is visible in the morning skies a week after Full Moon. Because the morning Half Moon has more dark 'seas' it means they are not as bright as the afternoon ones – and this adds to the reason why some humans don't notice the Moon during the day.

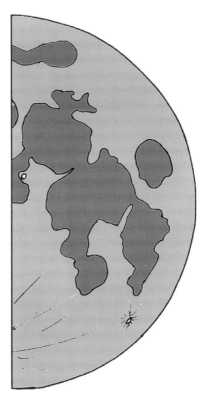

A morning gibbous Moon.

The Moon is easily visible in the daytime; it's just not so prominent due to the bright blue sky. In fact there is also a factor that makes a morning-sky, **waning moon** less bright – the positions of the dark 'seas' on its surface.

Depending on where you live on our planet, the Moon looks quite different, both in appearance and movement across the sky. Take the evening-crescent, **waxing moon** shown opposite, here complete with earthshine – the bright crescent lit by the Sun, the rest lit by the Earth. This is how the Moon looks at the same

time from different places on the Earth. So, if you're not used to it, the Moon can appear quite bizarre.

And it's a myth that the Moon is bigger near the horizon than high in the sky. It's an optical illusion – well, at least it looks like one.

A mid-northern latitude view of the evening crescent Moon.

An equatorial view of the same Moon looks like this...

...and in the mid-southern latitudes it looks like this.

For those who are interested, here is some basic Moon info:

Diameter: 3,475.5 km (or 2,159.6 miles, or 17,773 furlongs)

Mean distance from centre of the Earth: 384,400 km (or 238,855 miles, or 76.4 million rods)

Sidereal month (means it rotates on its axis in): 27.32 days*

Synodic month (described more fully on page 111, means its phases repeat every): 29.53 days

Orbital velocity (or mean speed around the Earth): 3,680 km/h (or almost exactly mach 3)

Mass: 73,500,000,000,000,000,000,000 kg (or 73.5 yottagrammes)

*Note that the sidereal month length is not the same as a synodic month length... (thinking time)... a single spin of the Moon is not the same as the time it takes to go from Full Moon to Full Moon... (thinking time)... If this is of no concern to you then please feel free to skip ahead to Eclipses (overleaf). For anyone else, just remember that as the Moon orbits the Earth, the Earth is also orbiting the Sun. Imagine for a moment that the Earth was stationary in its orbit – both sidereal time and synodic time would be the same. However, with an orbiting Earth, the Sun now appears to be in motion in relation to the background stars – in fact this is why and how the Sun moves slowly through the 12 signs of the zodiac. In about one month, the Sun has travelled almost 1/12th of the way around the sky, so in this time the Moon has to catch the Sun up – and it needs just over two days to do that, hence the difference. Don't you just love space?

9 January 2001 at 20.18. A total lunar eclipse is an amazing sight: this red object sits in a sky full of stars, looking very much like an alien world here for a visit. I took this picture from the Royal Observatory in Greenwich, during one of their eclipse events, and there is a little bit of light pollution around, as you can see. The further away you can get from street lights and big towns, the darker and better the skies will be.

Eclipses

Lunar eclipses

A total eclipse of the Moon is just one of three possible forms of lunar eclipse. The others are a partial and a penumbral eclipse, but they're not in any way as exciting as a total, so I won't mention them again. A total lunar eclipse can only occur at Full Moon, when the Sun, Earth and Moon are perfectly lined up in space. By looking down on our planet from over the North Pole (see the super eclipse diagram opposite) we can see what goes on.

From **position 1**, the Moon, being fully lit by the Sun, starts to travel into the shadow of the Earth. Over the course of a few hours, the

Moon approaches the exact opposite side of the sky to the Sun, where the Earth blocks out all direct light falling on it. Normally this would be at the time of a Full Moon, but now we're in for a treat due to the perfect alignment of the three celestial bodies.

Up in the sky you will have noticed the left side of the Moon gradually darkening over this time until totality is achieved at **position 2**. Strangely the Moon often takes on a reddish-orange-brown mix of colours and is very rarely black. This is because indirect sunlight is still reaching it – an effect of the Earth's atmosphere which lets the red part of the Sun's light filter through to dimly light the Moon.

Most months the Moon moves above or below the Earth's shadow, which is why lunar eclipses do not happen at each Full Moon. However, there is generally some sort of line-up at least a couple of times each year.

Top tips for lunar-eclipse viewing

As the Moon goes into eclipse and dims, the sky gets darker too. You may not have realised how the Full Moon causes the sky to be awash with a blue haze, allowing only the brighter stars to be visible. During a total lunar eclipse this darker Moon means the fainter stars can come out and we end up with an eerie sight of (usually) a deep red Moon surrounded by twinkling stars. Have a look; you'll see what I mean.

Totality can last for over an hour and a half, so the whole thing is quite a gentle process. You need no special equipment, and if your house is pointing in the right direction you don't even need to go outside! You can simply gaze spellbound out of your window. Now that's luxury.

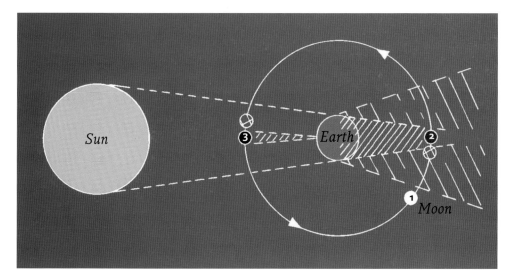

This eclipse diagram is not to scale but shows the principle. We need three things for an eclipse: the Sun, the Earth and the Moon (shown in various positions). In position 1 there is no eclipse. Only when the three things are lined up do we get an eclipse, for now the Earth's shadow can fall on the Moon (a lunar eclipse – at position 2) or the Moon's shadow can fall on the Earth (a solar eclipse – at position 3).

The evening partial
eclipse of 30 May
1984.

Solar eclipses

If a badger walks down the street on a sunny
day into the shadow of a building, you can say
that the badger sees the Sun being 'eclipsed'
by the building. The same thing happens with
much larger objects, such as the Moon, but of
course the shadow is much larger. In fact, the
Moon's shadow is nearly 3,500 km in diameter.
This happens when the Sun, Moon and Earth
are perfectly lined up – the time of New Moon
– as shown on the super eclipse diagram at
position 3 (see page 117).

However, because the Moon is much
smaller than the Sun, the shadow is actually
a cone shape (notice that on the diagram, too).
When the Moon is close enough to us in its

orbit, we find that its shadow cone can just
touch the Earth. This is the time, and place,
of a total solar eclipse, or a total eclipse of the
Sun. Therefore, this type of eclipse is only seen
on certain parts of the Earth.

As the Moon moves along its orbit, the
shadow moves across the Earth at a rate of
about 3,200 km/h (this is a very rough
approximation, as the speed changes
constantly – slowing down and speeding up as
the shadow moves over the curved surface of
our planet). The area covered is along the path
of totality, and only when you are standing on
this path will you see a total eclipse. Further
north or south of this line you will see a
partial eclipse, in which only part of the Sun
is covered.

Of course, the line-up can be good enough
to cause an eclipse, but not good enough for
the cone to fall on the Earth, in which case a
partial eclipse of the Sun is the best we can see.

An annular eclipse is caused by the elliptical
nature of the Moon's orbit, which over a month
takes it closer or further from us. When it is far
enough away, it appears smaller than the Sun,
so if everything happens to line up at this time
it is unable to cover the Sun completely (a total
eclipse) and we see a ring of sunlight
surrounding the Moon – an annular eclipse.
(The name comes from *annulus*, the Latin word
for 'ring'.)

Like their lunar equivalents, solar eclipses
do not happen every month, due to the fact that
the Moon's orbit around us is tilted about 5° to
the Earth's orbit around the Sun. This means
that the Moon's shadow usually passes above
or below our planet at New Moon. But at least

twice a year everything lines up so that some part of the Moon's shadow falls on the Earth's surface and a partial, annular or total eclipse of the Sun is seen from somewhere.

Viewing a solar eclipse

(including events seen only during a total eclipse)

FIRST CONTACT: This is the instant the Moon starts to move in front of the Sun. Very slowly you will notice an increasing 'bite' being taken out of the Sun.

Darkening skies: For about half an hour you will not really notice anything unless you know the eclipse is happening. This is because light levels are falling very gradually and the Sun's disc is bright enough to overcome the encroaching Moon.

Trees: Try to find a tree through which to watch the dappling of sunlight. Usually the pin-hole effect of the leaves causes circles of light on the ground. During an eclipse these dapples are transformed into hundreds of crescents.

Stars and planets: Just before totality the sky can be dark enough for the brighter stars and planets to be visible.

Plants and animals: Birds begin to fly back to their nests, and nocturnal animals may appear. Listen out for any owls and watch as some flowers begin to close their petals. The temperature is also dropping, and it can be quite chilly by the time of totality.

SECOND CONTACT: This is it! Totality begins. If you are quick enough, because there is so

First contact: slowly the Moon moves in front of the Sun.

Totality: 2 minutes and 30 seconds of wonder.

Third contact: the 'Diamond Ring' effect heralds an end of the total phase.

much going on, you may be able to witness the Moon's shadow rushing through the atmosphere towards you from the west, while at the same time the final part of the vanishing Sun is just visible through all the mountains and valleys around the edge of the Moon – this effect is known as 'Baily's Beads'. As far as totality is concerned, you will have from one second to seven and a half minutes to enjoy this spectacle. Totality is the only time when the Sun's outer atmosphere, the corona, is visible. This is a pearly-white, delicate structure of streamers weaving away from the Sun. Then, after too short a time, it's all over...

THIRD CONTACT: This is the moment that everybody claps and cheers – the Sun has just peeped out from behind the Moon and all the fuss is over. Because your eyes have become slightly dark-adapted during totality, the emerging Sun's light appears spectacularly, together with the glow around the Moon, all leading to what is known as the 'Diamond Ring' effect. Over the next hour and 20 minutes there is a rewind through the first-contact events as everything slowly returns to normal.

FOURTH CONTACT: The final 'bite' by the Moon disappears and the Sun is 'full' once again.

You'll only see all of these contacts if it's a total eclipse you're watching – partial eclipses go from first contact to fourth with varying degrees of Sun coverage in between.

Viewing solar eclipses *safely*

The Sun can only be viewed safely with the naked eye during the few brief seconds or minutes of totality. It is *never* safe to watch any partial view of the Sun without special precautions. Even when 99 per cent of the Sun's surface is obscured during the partial phases of a total eclipse, the remaining crescent is intensely bright and cannot be viewed directly without proper eye protection.

And I have to dispel a dangerous myth about watching solar eclipses: *never* watch an eclipse by looking into a reflection in a pool of water – the Sun is only slightly dimmed and can still do your eyes plenty of harm.

You can buy eclipse viewers, which cancel out all dangerous radiation and 99.9 per cent of the light. If you decide to use these, check to make sure they have the correct certification marks on them (the CE or British kite mark) and that they are not damaged in any way. Some experts advise never looking at the Sun, no matter what, but a certain amount of common sense is required on your part. There are those who believe that even a glance at the Sun causes irreparable damage. I think several million years of evolution may have something to say about that.

> **Final warning**: Do not look at the Sun with the unaided eye. Failure to use proper eye protection may result in permanent eye damage or blindness.

Step-by-step guide to pin-hole eclipse watching

The simplest and safest method for watching a partial or annular eclipse (or the partial phase of a total eclipse) involves nothing more complicated than two bits of card. Make a small pin-hole in one and allow the light from the Sun to fall through it on to the other. It's that simple! The hole will produce the exact image of the Moon passing in front of the Sun as the eclipse happens. Remember, *never* look through the pin-hole itself at the Sun. If you find the image is not great, make sure the hole is as round as possible or try a slightly larger or smaller hole – use your initiative for making a hole smaller!

The Planets

Orbiting the Sun there are nine known major planets: Mercury, Venus, Earth and Mars are relatively small and rocky, while Jupiter, Saturn, Uranus and Neptune are much larger and gaseous. Tagged on to the end is the tiny, frozen, semi-planetary world Pluto, which I have no problems in classing as a planet.

Even further from the Sun we know that there are more objects, some of which could be larger than Pluto. If one is found in these dark depths then no doubt it will be called a new planet – the mythical planet 'X' (Roman numeral for 'ten'). A new icy world, named Sedna, discovered on 14 November 2003, was a possible planetary candidate for a while. Unfortunately, after calculations, it turned out to be only half the size of Pluto and so it is classed as a 'minor planet' or 'asteroid'. However, I have every confidence that as telescopes and detection methods improve we will get another planet for our solar system.

How did this marvellous group of planets come about? Well, in our space neighbourhood about 5 billion years ago, we would have seen a very stormy picture indeed. Parts of an immense cloud of dust and gas (our solar nebula) were clumping together under gravity, producing vast amounts of heat and energy as they went. One of these clumps was to become the Sun. An amazing thing about a 'clump' is that the more stuff you add to it the more gravity it has, so as the Sun was coming into existence, its increasing gravitational force meant that a disc of dusty material that was to

Do not try this at home. I know it was crazy, but look what I did one night while nobody was looking – I arranged all the planets so you could see exactly how they differed in size. Thankfully I had a handy gravity device to keep them from becoming one fine mess. If they were really this close without my gadget in operation there would be all sorts of gravitational tidal forces to contend with. Anyway, it is not difficult to see that Jupiter is the largest of the planets; so large, in fact, that you could squash 1,300 Earths inside. Behind, Saturn stands out due to the finest planetary rings of the solar system – these really make the ring systems that we find around Jupiter, Uranus and Neptune pale into the insignificance. (Image courtesy Hubble/AURA/STSci/ NASA/JPL/Greg Smye-Rumsby)

The two-part solar system: inner rocky versus outer gassy. Not only that, but the distances are such that a diagram large enough to explain things fully needs to be split into two. The inner worlds are so much closer to the Sun, and as we travel out the distances get 'astronomical'! (Image courtesy Hubble/AURA/STSci/NASA/JPL/Greg Smye-Rumsby)

become the planets was also forming.

About 4.7 billion years ago, once the temperature within our clump reached about 10 million degrees Celsius – the magic point at which nuclear furnaces switch on – the Sun finally began its 10-billion-year life cycle and caused a shock wave of radiation that blasted through the happily growing planetary disc. Most of the lighter gases that were hanging around were blown far from the new sun – which is why we find the gas giants in the outer solar system. All the heavier hard material was able to withstand the blast and stay where it was, hence the rocky inner planets.

Taking everything into account, we know that the Sun is now around halfway through its life – so we have only 5 billion years in which to build a spaceship and find a new home before the Sun swells up and turns into a red giant and the Earth becomes fried, crispy and dry.

Meanwhile, there are lots of ways to remember the names of the planets that we do have. How about 'Several Murky Vans Entered Margate at Junction Seventeen Unfortunately Needing Petrol' – the first letters of each word

are the first letters of each planet, plus the Sun. However, I've also heard, 'My Very Elderly Mother Just Sips Unrefined Nettle Pulp'! Charming. Use whatever works for you.

One of the best ways to look at how everything fits together is by judging each planet's distance from the Sun in relation to that of the Earth. Our average distance from the Sun is known as 1 astronomical unit (or AU). Now compare our neighbours: Mercury 0.39 AU, Venus 0.7, Mars 1.5, Jupiter 5.2, Saturn 9.5, Uranus 19.2, Neptune 30.1 and Pluto out at a whopping 39.5 AU. Notice how the inner planetary numbers are relatively close to Earth's 1 AU while as we move outwards in the solar system the numbers grow at a tremendous pace. Three important things happen as we travel out: 1) The shrinking appearance of the Sun means that each planet gets less heat. This leads to a surface temperature on the sunny side of Mercury of 350°C, while distant Pluto freezes at −230°C. 2) For the same reason there is less sunlight, and by the time you reach Pluto the Sun really does just look like a bright star. 3) Due to the decreasing gravitational tug of the Sun, the speed at which a planet orbits decreases with distance. Of course, the outer planets have further to travel – like the outside lane of a running track compared to the inside track – and so we find Mercury orbits the Sun in just 88 days while Pluto takes 248 years!

Forgetting the solar system for a moment, the astronomical unit can be used instead of light-years to understand (with some thought!) how far away the nearest star, Proxima Centauri, is. And we find it is 268,710 AU. Can you imagine that?

Mercury

This innermost world moves the quickest, orbiting our star four times faster than the Earth. The great heat from the Sun here doesn't allow any atmosphere to exist on Mercury, and without this control system the daytime temperatures can reach 400°C, whilst at night the thermometer plunges to -170°C. How would you like your body? Fried to a crisp or frozen till bits start falling off? Mercury isn't a viable holiday destination.

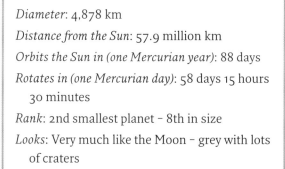

Diameter: 4,878 km

Distance from the Sun: 57.9 million km

Orbits the Sun in (one Mercurian year): 88 days

Rotates in (one Mercurian day): 58 days 15 hours 30 minutes

Rank: 2nd smallest planet - 8th in size

Looks: Very much like the Moon - grey with lots of craters

OBSERVING MERCURY: Mercury is quite a small planet and the word 'elusive' is often used in terms of trying to find it. This is because this innermost planet stays close to the Sun and so never appears far above the twilight horizon. Believe it or not, even some astronomers have not seen Mercury! However, it can be found fairly easily if you know where and exactly when to look. Due to various tilts out there in space, the best times to see Mercury are the evening skies of northern hemisphere spring/southern hemisphere autumn, or the morning skies during northern autumn/ southern spring.

24 May 2001 at 21.00. I've seen Mercury on plenty of occasions, but this image, taken in Alcudia, Mallorca, was the first time I'd spied it above the Moon.

7 May 2003 at 08.30. Mercury occasionally crosses the Sun's disc (a transit) – something to watch only using extremely safe observing methods (as for a solar eclipse, as described on pages 120–121). On this picture the big splodge to the lower left is a sunspot (a place where the Sun is a bit cooler) while the lone black dot just above the centre is Mercury. Even though they look as if they're at the same distance, Mercury is still nearly 58 million km from the Sun, which just gives you some idea of the vastness of space.

4 May 2002 at 21.23. It's a fine evening in Kemsing, Kent, and the two planets Mercury and Venus are visible in the dusk. Using our imaginations we can magically draw in the orbits of these worlds. Notice that Venus can, and will, travel in the direction of the arrows much further above the horizon than Mercury. In fact, looking at Mercury you can see it is already at the furthest visible point from the Sun, meaning that it is never visible in a really dark sky, and also it is only around for a short time after sunset before it too is carried below the horizon.

A mosaic of Mercury made up of eighteen images from the first Mariner 10 spacecraft flyby in March 1974. (Image courtesy NASA)

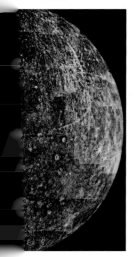

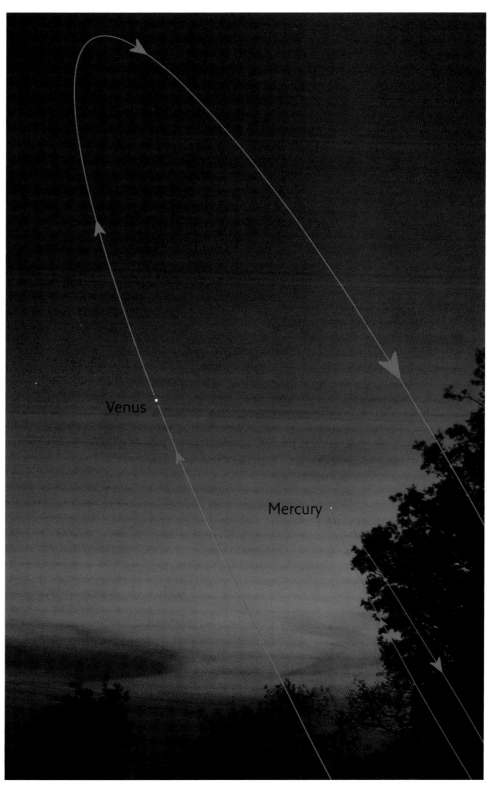

Venus

Mercury

8 June 2004. It was a Tuesday. For the first time in 121.5 years Venus could be seen crossing the Sun. Historically, this was the most watched transit of Venus ever. The size of the planet over the Sun was staggering compared to the tiny speck that was Mercury when it was in transit the year before. As one person commented, 'It's like someone hole-punched the Sun.'

The Mariner 10 spacecraft snapped this picture of cloudy Venus back in February 1973 while on its way to Mercury. In fact this craft for the first-time ever used a technique used now for all deep-space probes called 'gravity-assist'. This basically involves using the gravity of a planet to change course and speed to reach your required destination – a sling-shot effect.

Venus

Planet two orbits the Sun in a shorter time than it rotates, meaning that a Venusian day is longer than its year! It also comes closer to the Earth than any other planet at only 40.5 million kilometres, which is just 100 times further away than the Moon.

> *Diameter*: 12,104 km
>
> *Distance from the Sun*: 108.2 million kilometres
>
> *Orbits the Sun in (one Venusian year)*: 224.7 days
>
> *Rotates in (one Venusian day)*: 243 days 0 hours 30 minutes
>
> *Rank*: 6th in size
>
> *Looks*: Today's weather: cloudy

OBSERVING VENUS: Venus can be the brightest object in the sky after the Sun and Moon. Sometimes this means it can even be seen in full daylight while at night it can cast shadows. This brightness is caused by nice fluffy white clouds of deadly carbon dioxide which bounce off about 65 per cent of the sunlight hitting them; also Venus comes closer to the Earth than any other planet. Not surprisingly the ancients called it the Evening or Morning Star, depending on when it was visible – but visible it most certainly was.

Very rarely Venus can be seen, with precautions, crossing in front of the Sun. These so-called *transits* occur in pairs separated by

28 December 2000 at 16.57. Venus and the Moon in the evening skies can be a great sight, especially when Venus is shining at its brightest.

over one hundred years. The last one was on 8 June 2004 and will be followed by one on 6 June 2012. Miss that and you have to wait until 11 December 2117!

29 December 2000 at 16.50. This image of our two nearest neighbours, the Moon and Venus, was taken one day after the image above and clearly shows how the Moon moves from day to day. On the following evening the crescent Moon was above and to the left of the planet – I would have taken another picture but, as usual, it was cloudy.

Earth as seen by Apollo 17 on 7th December 1972. This was the first time the south pole had ever been seen from space. (Image courtesy NASA)

A gibbous Earth picture taken from lunar orbit. (Image courtesy NASA.)

To date, only 27 humans have seen an Earth-phase (in this case a half-Earth) – that is, those who have gone to the Moon. This image came from the Apollo 10 command module, named Charlie Brown, with the lunar-lander, named Snoopy, pictured whilst in lunar orbit in May 1969. However, there is a possibility that NASA conducted several secret lunar missions in the early 1970s, which means there are more people around who have witnessed such an amazing sight as this. The Earth travels along a unique orbit – if we were closer to the Sun, we would be too hot, but any further away and we would be too cold. This orbit is the so-called 'habitable zone', which has meant that we have a warm friendly atmosphere and liquid water, all allowing the creation of life. (Image courtesy NASA)

Earth

Our planet orbits at a distance from the Sun that would take you 2,123 years to walk or 193 years to drive at 90 km/h. We have one natural satellite, the Moon, which is very slowly spiralling away from us at a rate of 1.5 km every 28,000 years. This means that it will appear smaller in the future, to the point where total solar eclipses will no longer occur as the Moon will not be large enough to cover the Sun's disc – a sad day indeed if you don't have a rocket.

Some 78 per cent of our atmosphere is the gas nitrogen, with oxygen making up 21 per cent.

Diameter: 12,756 km

Distance from the Sun: 149.6 million km

Orbits the Sun in (one Earth year): 365.25 days

Rotates in (one Earth day): 23 hours 56 minutes 4 seconds

Rank: 5th in size

Looks: 70 per cent covered in water, so quite a bluish world

OBSERVING EARTH: Look up, look down, look all around... and wonder.

Mars

This has been an endless source of fascination due to its striking red colour, astronomers' drawings of canals on the surface, H. G. Wells's *War of the Worlds* and more recently the search for the 'lost' oceans of Mars.

In 1994 there was a much reported study of the groovily named meteorite ALH 84001 that was found in Antarctica. According to some it had come from Mars carrying fossilised bacterial life. Other reports since then, however, have cast doubt on so-called 'evidence' of this Martian life. With all the current missions to Mars, hopefully one day we will know the true story of whether there was or is life on our red planetary neighbour.

Mars has a thin atmosphere with winds that pick up the red rusty Martian dust, carrying it like sand storms across the planet.

OBSERVING MARS: Mars can approach the Earth as close as 55.7 million km or move as far away as 400 million km. There is also the effect of the oval nature of Mars's orbit to consider. Indeed, on 27 August 2003 the orbit was such that Mars was at its closest to us for almost 60,000 years! This made it an extremely bright object. More generally the Earth catches up and passes Mars every 18 months or so, the time when this red world can become the second brightest planet in the sky (after Venus).

Diameter: 6,787 km

Distance from the Sun: 227.9 million km

Orbits the Sun in (one Martian year): 686.9 days

Rotates in (one Martian day): 24 hours 37 minutes 23 seconds

Rank: 7th in size

Looks: Red and rusty

Here's a drawing I made of Mars on 13 May 1999 at 21.55 GMT using my 80 mm refractor. The dark V-shaped feature is Syrtis Major, an easy feature to see (as are many others) when Mars and the Earth are close together in their orbits.

Many great pictures of planets are mosaics, and this is no exception. Nearly 100 Viking mission images were used during the Martian summer of 1980 to produce this global view of Mars as seen from 2,500 km up. In the centre is the large 470 km diameter Schiaparelli crater, while the white area to the lower-right is frost in the Hellas Basin. (Image courtesy NASA)

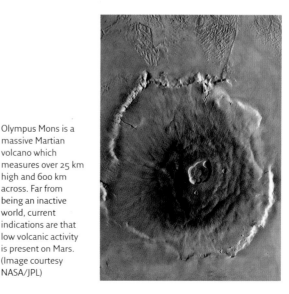

Olympus Mons is a massive Martian volcano which measures over 25 km high and 600 km across. Far from being an inactive world, current indications are that low volcanic activity is present on Mars. (Image courtesy NASA/JPL)

The Asteroids

Between the orbits of Mars and Jupiter are loads of space rocks known collectively as the Minor Planets – this is the Main Asteroid Belt. One theory about their formation is that a planet was unable to form here because of the massive gravitational influence of nearby Jupiter.

Ceres is the largest in the main belt, measuring 940 km (584 miles) in diameter, and was also the first to be discovered in 1801. Then came Pallas, Juno and the brightest asteroid, Vesta. Several of these minor planets have quite down-to-earth names such as Hilda, Albert and Thora. Some are even named after rock stars, including Enya, (Eric) Clapton, (Frank) Zappa and (Jean-Michel) Jarre.

OBSERVING ASTEROIDS: There is one that you can see at times with the unaided eye: welcome on stage – Vesta. It appears only as a faint star, so you'll need good dark skies, but the challenge is there.

Eros is a wonderfully shaped rock measuring 33 x 13 x 13 km. On Monday 12th February 2001 it became the first asteroid in history to have a robotic visitor from Earth – the NEAR (Near Earth Asteroid Rendezvous) spacecraft landed after completing just over one year in orbit closely studying this asteroid. (Image courtesy NASA)

Ida was captured in this image by the Galileo spacecraft on Saturday 28th August 1993 as it flew past on the way to Jupiter. This asteroid, measuring 56 x 42 x 21 km, was the first found to have its own moon, a small 1.4-ish km round pebble named Dactyl. (Image courtesy NASA)

Jupiter

Here is the largest planet in the solar system, and the first of the gas giants. Who knows how many moons Jupiter has? The intense gravity of the planet could mean that there are hundreds of the things! Most are incredibly small and we may never know the real number. Jupiter also has the famous Great Red Spot, a storm that has raged for over 300 years. It is so big you could fit two Earths inside it. (By the way, Jove was another Roman name for the god Jupiter, which is why the adjective is *Jovian*.)

Diameter: 142,800 km

Distance from the Sun: 778.3 million km

Orbits the Sun in (one Jovian year): 11.86 Earth years

Rotates in (one Jovian day): 9 hours 50 minutes 30 seconds

Rank: The big one! Numero uno

Looks: Big, big, big

OBSERVING JUPITER: A big world that reflects a lot of sunlight, hence at times Jupiter appears as a very bright star indeed. You'll need a telescope to see the famous atmospheric belts and the Great Red Spot, and simple binoculars to see the four tiny dots that are the main moons.

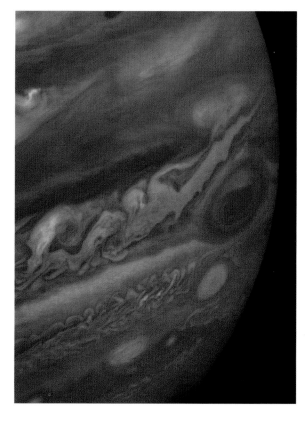

A Voyager 2 image of Jupiter from 9 million km taken on 29th June 1979. (Image courtesy NASA/JPL)

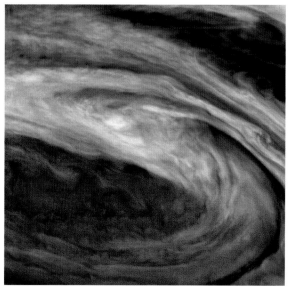

Here is a Voyager 1 spacecraft close-up image showing the intricate details of the Great Red Spot storm on Jupiter during February 1979. (Image courtesy NASA/JPL)

Saturn

Give it up for the second largest planet – the one with the rings! Actually all the four gas-giant planets have rings – Jupiter, Saturn Uranus and Neptune – it's just that the bits that make Saturn's are brighter and there are more of them. Saturn is extremely light due to its gaseous make-up, and given a large enough bath and a good supply of water, you would find that this planet would float!

The rings themselves are made up of particles of icy rock from as small as a grain of sand up to the size of a house, each individually orbiting like tiny moons.

Diameter: 120,000 km

Distance from the Sun: 1,427 million km

Orbits the Sun in (one Saturnian year): 29.46 Earth years

Rotates in (one Saturnian day): 10 hours 14 minutes

Rank: Numero Dos

Looks: Probably the best of the bunch

OBSERVING SATURN: Like Jupiter, Saturn is a pretty big world that can appear quite bright when it and the Earth are in the right place. You really need a small telescope to see the rings and moons (including Titan), as binoculars are just not powerful enough.

A marvellous view of Saturn composed of 126 glued-together images taken over two hours by the Cassini spacecraft on 6th October 2004. Image courtesy NASA/JPL

Titan, Saturn's largest moon, as seen by the Cassini spacecraft on 15th February 2005 from a distance of 229,000 kilometres. This is basically the view you would see out of a spaceship window – a small world with a chemically hazy atmosphere. Image Courtesy NASA/JPL/ESA

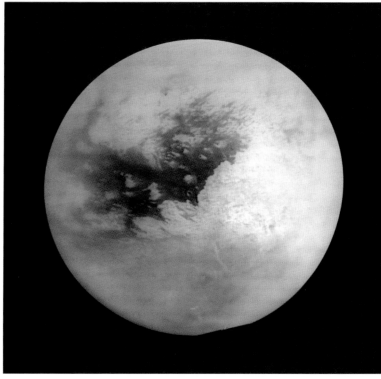

Now with special computery techniques the haze seen in the previous picture can be removed to reveal a never-before-seen view of Titan's surface. In fact this is a mosaic of 16 images captured by the Cassini spacecraft. Image Courtesy NASA/JPL/ESA

Uranus

This planet was the first to be discovered using a telescope. The honour goes to William Herschel on 13 March 1781 – although loads of people had seen it before that, nobody realised what it was. Herschel originally named this new world 'Georgium Sidus' in honour of King George III, but Uranus was eventually adopted because it was a more classical name. The most unusual aspect of Uranus is that its axis is tilted so much that the planet appears to spin on its side, similar to the way a ball rolls along the ground.

The Voyager II spacecraft catches a final view of Uranus on 25th January 1986 as it heads off to Neptune. This picture is much more interesting than a completely lit view of the planet, as Uranus shows little more than a featureless bluey/green ball to the eye. Image courtesy NASA/JPL

Diameter: 51,118 km

Distance from the Sun: 2,871 million km

Orbits the Sun in (one Uranian year): 84.01 Earth years

Rotates in (one Uranian day): 17 hours 55 minutes

Rank: 3rd

Looks: Nothing more than an uninteresting greenish fuzzy blob

OBSERVING URANUS: Uranus is just visible to the unaided eye when at its brightest, which is magnitude 5.5. This is challenging indeed, even for those with the super-clear, dark starry skies needed for this task.

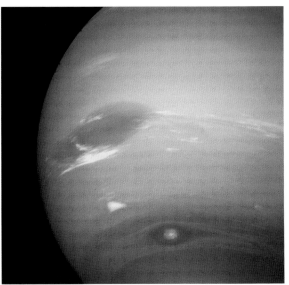

This is a picture of Neptune taken on 24th August 1989 by the Voyager II spacecraft. It shows the stormy nature of this gassy planet at that time: the large feature to the middle-left is the Great Dark Spot. Image courtesy NASA/JPL

Neptune

The last and smallest of the four gas-ball giants, Neptune is still 54 times bigger than the Earth. Due to its great distance from the Sun, Neptune is quite a faint world, sitting as it does near the cold lonely edges of our solar system. So it shouldn't come as any surprise to learn that this planet was not identified until 1846 – although Galileo may have seen it in 1612.

At times, thanks to Pluto's unusual orbit, Neptune can be the furthest planet from the Sun.

Diameter: 49,528 km
Distance from the Sun: 4,497 million km
Orbits the Sun in (one Neptunian year):
   164.79 Earth years
Rotates in (one Neptunian day): 19 hours 10
   minutes
Rank: 4th
Looks: Sad and blue

OBSERVING NEPTUNE: Because of Neptune's great distance from the Sun, you need binoculars to find this 'star', as it only reaches magnitude 7.7.

Pluto

Cold Pluto was named after the god of the underworld because of its tremendous distance from the Sun. This is the smallest of all the planets, smaller than our Moon, in fact, which together with its remoteness explains why it wasn't discovered until 1930.

Pluto takes 248.54 years to go round the Sun in a highly eccentric orbit that brings it inside the orbit of Neptune for 20 years of its circuit (the most recent was 1979 to 1999).

Diameter: 2,320 km
Distance from the Sun: 5,914 million km
Orbits the Sun in (one Plutonian year):
   248.54 Earth years
Rotates in (one Plutonian day): 6.387 days
Rank: Tiniest
Looks: Perhaps like a dog?

OBSERVING PLUTO: Pluto is so distant that only mega telescopes will be able to find its faint (magnitude 13.8) dot in the heavens. And don't bother trying this at home if you live in any sort of light-polluted area.

As yet, Pluto is the only planet not to have been visited by a spacecraft. This image was produced by the Hubble Space Telescope – which is in orbit around the Earth. Amazingly from this vast 6 billion-kilometre distance, light and dark patches can easily be made out on this small distant world. Image courtesy NASA/ STScI/Hubble

Planets and Days

Mercury, Venus, Mars, Jupiter and Saturn are known as the ancient planets, as all of our long-gone civilisations could see them with the unaided eye. Together with the Sun and the Moon this made a total of seven objects that moved about the unchanging starry skies. Not surprisingly, especially considering that it was believed all the gods lived 'up there', seven became a very important number, and indeed this is why we have seven days of the week.

While cultures such as the Greek, and others further to the east, simply numbered most of these seven week days, many western European languages integrated the planets even further by naming the days after them.

This table shows the connection between day names and those objects that moved around the classical heavenly skies. There have been some changes, but it's clear where it all came from:

English days	Gods for English days	Equivalent Roman gods (planet names)	Greek gods	Latin	French	Spanish	Welsh
Saturday	Saturn	Saturn	*Kronos*	Saturnus	*Samedi*	*Sábado*	Dydd Sadwm
Sunday	Sun	Sun	*Helios*	Sol	*Dimanche*	*Domingo*	Dydd Sul
Monday	Moon	Moon	*Selene*	Luna	Lundi	Lunes	Dydd Llun
Tuesday	*Tiw*	Mars	*Ares*	Mars	Mardi	Martes	Dydd Mawrth
Wednesday	*Woden*	Mercury	*Hermes*	Mercurius	Mercredi	Miércoles	Dydd Mercher
Thursday	*Thor*	Jupiter	*Zeus*	Iuppiter	Jeudi	Jueves	Dydd Iau
Friday	*Freya*	Venus	*Aphrodite*	Venus	Vendredi	Viernes	Dydd Gwener

Notes

The *italicised* days are those with no apparent solar system connection in English. This, of course, includes Tuesday to Friday, which are named after Norse gods. However, digging deeper, we find that these are basically the equivalent of the Roman planet gods – an example is Freya, the Norse god of love. In English, Friday is named after her, while in French and Spanish the same day takes its name from Freya's Roman counterpart, Venus.

The French word for the Moon is *la lune*, while the Spanish have *la luna* – these link very closely with the Latin name. The Welsh names are extraordinary in that they show the influence of Roman Britain, and even more extraordinary considering that the language has no direct relation whatsoever to Latin.

The classical arrangement of the planets out from Earth – our planet, of course, considered to be the centre of the Universe – was: the Moon, Mercury, Venus, the Sun, Mars, Jupiter and then Saturn. This can be seen within our week by a curious jumping of days. Start from Monday, the Moon, jump over one day to Wednesday, originally Mercury. Continue this one day jump and you will obtain the list just given.

The Milky Way

Watching a dark sky during evenings from August to December in the northern hemisphere, or April to September in the southern, you'll have the best views of a faint band of light stretching across the heavens: this is the Milky Way.

This name comes from the time of the Greeks, who, as we have seen, believed the Earth was fixed in the centre of the Universe with the Sun, Moon and planets orbiting around us. Beyond all this was a crystal sphere to which the stars were attached. The story goes that on one fateful night (using Roman names) Juno, the wife of Jupiter and someone who liked to be in charge, spilt her milk across this starry globe – yes indeed, this was to make the Milky Way.

It wasn't until Galileo (probably) first looked through a telescope that this 'fuzzy band' was revealed to be not milk, but thousands upon thousands of stars. Astronomer William Herschel calculated correctly that these stars took the form of a large disc, which we are

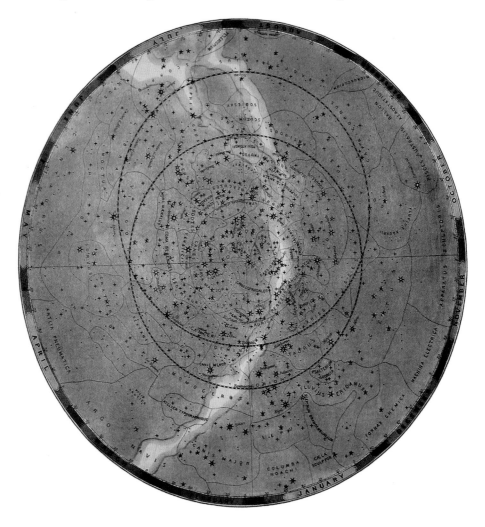

The Milky Way as depicted on a 19th-century design of the northern skies.

We see galaxy NGC 4013 as an edge-on spiral. If we were to fly straight into this galaxy we would see the majority of stars and dust in a band all around us. In exactly the same way, the Milky Way appears as a band across our night sky. (A Hubble image courtesy AURA/STScI/NASA)

inside. From our position we can look out along the plane of this disc to see all the other disc stars around us forming the faint Milky Way band. Imagine lots of people sitting in a field, all on the same level with you in the middle, so everyone's head (representing the stars) appears more or less as a line around you – all on the same plane.

So, all the stars we see at night, including the Milky Way, are just a tiny part of a vast disc-shaped starry island that we call the **Galaxy** – our home in the Universe.

It took radio astronomy (which uses those big dish-shaped telescopes) to discover the spiral structure of our Galaxy – looking from above like a rotating Catherine-wheel firework. Not spinning as fast, of course: out where we are the Galaxy takes about 225 million years to rotate once – in astronomical terms this is known as a *cosmic year*. Okay, not a very

exciting firework display, but then there's a lot to rotate – over 200,000 million stars stretching more than 100,000 light-years across.

With the vast Catherine wheel in mind we can visualise each spiral of stars emanating from the centre. Each spiral is known as a *galactic arm*, containing millions of stars together with plenty of dust and gas. Our Sun, and therefore our planet, are located near the edge of the so-called Orion arm (or 'Orion spur' as it is sometimes known as we're not sure if it is a real arm or just a broken section) about 30,000 light-years out from the galactic centre.

Looking out from our galactic home we see that the rest of our Galaxy is at its brightest if we gaze towards the constellations of Sagittarius and Scorpius, and faintest as we look out along the plane of the Galaxy towards Orion's club.

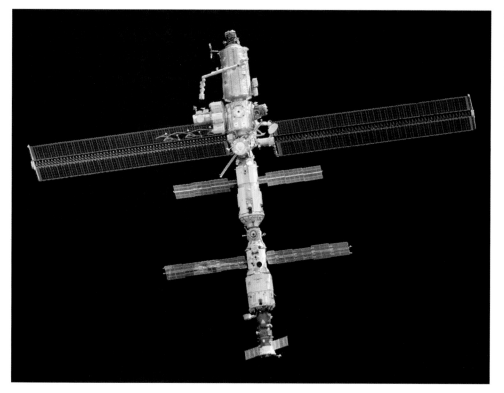

Watching Satellites and the ISS

Is it a bird? Is it a plane? No, it's the International Space Station. This slowly growing piece of space hardware appears in the night sky like a star that gently wanders from west to east over a few minutes.

We've been able to see the station, known as the ISS, since the first bit went up on 20 November 1998. This was the Russian Zayra ('Sunrise') module. Since then many more sections have been added, each of which makes the ISS brighter and therefore easier to see.

So when can you view the station? In these grand modern times you can find out

everything you need to know from the wibbly-wobbly-web. NASA and others have internet sites where you tap in details of where you live and up pop the exact times, dates and where to look (see pages 154–5 for lots of great starry websites).

The only things that will really be useful to know are the directions north, east, south and west from your viewing position. Knowing the Sun rises in the eastern part of the sky and sets in the western part is a good start.

When the shuttle docks with the ISS, it first has to approach the station. You can watch this

17 Feb 2000 at 18.57.
On the right the
space shuttle Atlantis
moves ahead of the
International Space
Station.

manoeuvre as day by day two slow-moving flying dots appear to be closing in on one another. Then, one night, two become one for however long before 'undocking' and growing apart again.

But that's not all there is to see: there are plenty of other satellites that cross the starry skies – four at once is my top score. You may think you're looking at aeroplanes, but without sounds or flashing lights there's only one thing they can be. Their slow-moving appearance is deceptive, too – most are travelling around the Earth at about 27,000 km per hour and are quite a few hundred kilometres up from the ground. Within these parameters they orbit the Earth every 90 minutes.

As with the ISS, we see a satellite because, and only because, it reflects the light from the Sun. Of course there is nothing to reflect if the light goes out, and this is what happens when the satellite enters the Earth's shadow. Therefore you can probably expect an object orbiting us to disappear at some stage, or indeed appear at any time, anywhere in the sky.

So you think you can only see satellites at night? Not true! There are wonderful craft called iridium satellites that you can watch during the day in a clear blue sky. You will be amazed how bright these 'flare' as their solar panels, acting like a perfect mirror (only the size of a door!), reflect the Sun down to your exact position. If you get details of an iridium flare, take note that you must be within 2 to 3 km of any prediction, otherwise you will not see anything – space really is that precise.

Here is a small sample of objects you can see orbiting the Earth, just to give you some idea of how high they fly:

	Kilometres up	
	MIN.	MAX.
Hubble Space Telescope	580	596
International Space Station	382	393
Cosmos 1143	404	406
Cosmos 2369	830	851

Because of the nature of its orbit, you unfortunately cannot see the International Space Station if you live above 70°N or 70°S.

A truly marvellous
sight: here's how
Hale-Bopp looked
from my back garden
back in April 1997.

Comets

There's something out there in the depths of space. Something that is starting a journey towards the inner solar system; moving slowly at first, but increasing little by little until its fiery passage through the heavens makes it visible to us here on our tiny planet.

This 'something' is a comet. In fact, there are loads of them – they are one of the most numerous and unpredictable of all the bodies zooming through space. Hale-Bopp was the finest comet of recent times, clearly visible in the evening skies of April and May 1997.

Of course, the most famous is Halley's comet, named after the 17th-century astronomer Sir Edmund Halley. No other comet has been seen by so many people throughout history and the world – it was first recorded in 467 BC. Probably its most remembered return was during the Battle of Hastings when it was woven into the famous Bayeux tapestry, commemorating William of Normandy's invasion in 1066.

Comets are regular visitors to our skies and have been chronicled through the ages by all

Depictions of comets in medieval times often showed them as swords. This is because they were taken to be messengers of doom. But enough of the dark stuff – they're only 'hairy stars', after all!

cultures, but even the greatest thinkers had no idea what they were. In ancient Greece, Aristotle claimed they were hot dry outpourings from the ground that were carried across the sky. When they got hot they simply caught fire, rapid burning leading to shooting stars and slow burning leading to comets. Very nice, but not strictly accurate in any sense of the word. Almost 2,000 years later Galileo's theory proved no better – he thought comets were caused by sunlight refracting through the Earth's atmosphere.

For the correct starting point we have to go back to Edmund Halley. His interest in comets started after seeing one in 1678, and he set off to find out all he could about their recorded appearances from writings around the world. Sir Isaac Newton's principles of gravitation – that's him with the apple – had just been published, and Halley used these for his research. Before long he found that certain comets seemed to have the same orbit and that

the dates when they became visible were 76 years apart. Could these separate observations in fact be of the same comet? Halley thought so and he predicted that this comet would reappear in 1758. It did, and the rest, as they say, is history.

From the work of Halley we learned that some comets moved in orbits like the planets, but it wasn't until the 1950s that we finally had an idea as to what a comet was made of. Astronomer of the time Fred Whipple put forward a theory that the nucleus was no more than a 'dirty snowball' about 10 km in diameter. This wasn't universally accepted, but the return of Halley's Comet in 1986 cleared the matter up once and for all. A fleet of spacecraft was sent to intercept the comet at various distances. The closest approach was made by the European Space Agency (ESA)'s Giotto space probe, which passed only 600 km from the nucleus on 14 March 1986. Giotto confirmed the snowball theory.

The great comet of 1456 over Constantinople.

The six-tailed bright comet of 1744.

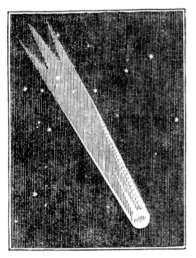

An etching of Halley's comet from the 19th century.

However, the comet turned out to look more like a large uneven potato than the round shape that had been expected.

As a comet-snowball approaches the Sun, the outer layers of ice from the nucleus vaporise, dislodging the dust which provides material for the comet's coma – that's the halo of gas forming the bright head of the comet. Sunlight then pushes this halo into the bright dust tail that we can see. Much fainter is the blue gas or plasma tail, which is caused by (boffin alert!!) the magnetic fields of the solar wind – a plasma flowing from the Sun at speeds of 400–720 km per second.

Comets are believed to be remnants from the formation of our solar system. When the Sun switched itself on, it blew all the light material to far beyond the orbit of the last planet Pluto into a halo known as the Oort cloud, or so another theory goes. Here lie over 100 billion comets, each waiting for a nudge from gravity which will send it falling toward the Sun. After a trip reaching speeds of up to 1.5 million kilometres per hour as it whips around the Sun, it may be gravitationally influenced again to become a periodic comet that returns to our skies at regular dates. On the other hand, it may fly off back into the deep space to rejoin its friends in the Oort cloud.

12 Aug 2002. A lone meteor on an atmospheric quest vaporises around 80 km above my garden. With an 'Ooh!' and a 'Woooow!' the neighbours Lee and Debbie are clearly as impressed as I am.

One of the first meteor storms that people took notice of was that of the 1799 Leonids. This engraving depicts the event as seen from Greenland.

Shooting Stars

What with planets, moons, asteroids, satellites and space stations, the solar system is a busy place. Then those comets, just mentioned, have tiny particles associated with them which, despite being the size of a grain of sand, generate the most spectacular sights in the night sky. If they happen to come into contact with the Earth's atmosphere, they burn up and we see the resulting trail as a streak across the sky known as a *shooting star* or **meteor**.

Many of you will have been out on a dark and clear night (some of you, I believe, only come out then, but that's another story) and made a wish after seeing a shooting star believing it to be a rare occurrence. In fact, these small grains are hitting the atmosphere all the time. There is plenty of this material hurtling through space, so if you gaze up into a clear night sky for any length of time, you should be rewarded with the sight of a shooting star.

The tiny particles are known as *meteoroids* while they fly around space. They hit our upper atmosphere, then becoming meteors, at anything up to 74 km per second. The trail we see is between 80 and 160 km above our heads and usually lasts rather less than a second.

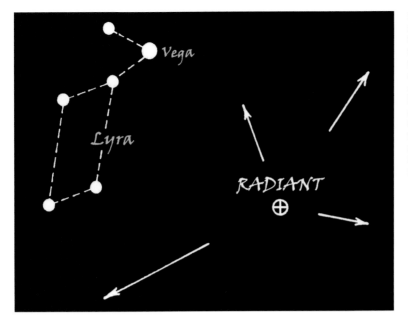

If you are watching a meteor shower, then the area in the sky where the shooting stars seem to emanate from is called the radiant. Hardly any meteors actually appear from this place; it's just that if you trace all shower trails backwards; you'll end up here.

So how many can you expect to see each night? Well, with clear skies and a low horizon, an average number is five meteors per hour. As I mentioned before, there are many grains hitting the atmosphere all the time, up to 100 million each day in fact, but the majority are too small to cause a meteor you can see. Of course, many arrive during daytime when there is virtually no chance of seeing them anyway.

Some of these streaks are caused by particles just aimlessly flying through space, then along came the Earth and *smash!* – a meteor. The name we give to these is *sporadic meteors*, because we cannot predict when they will occur. But there are swarms of particles out there orbiting the Sun that give rise to annual *meteor showers* which we can predict.

During this time we can see a big increase in the number of meteors each hour. These showers also appear from the same point in the sky, known as the **radiant**. Simply put,

each shower takes its name from the constellation where the radiant is located. For instance, in April there is a radiant in the constellation of Lyra, so the shower is called the Lyrids.

All in all, meteor showers allow you to see more meteors, plus you know when they will occur and where they'll be coming from. Most handy – thank you, space.

Meteor showers are where the association with comets enters the picture. For as comets fly through space, they leave a trail of debris orbiting the Sun – these are the meteoroids; as the Earth enters this debris, a meteor shower results. This was worked out by using observers at different locations and radar to calculate the orbits of these meteors before they hit the Earth. The results, because of a resemblance of these orbits to periodic comets, strongly suggest that meteoroids had their origins in the gradual decay of such comets.

The most powerful display in recent history came from the Leonids (from the constellation of Leo), on 17 November 1966, when over 2,000 meteors per minute occurred at the peak of activity. This shower is associated with Comet Temple-Tuttle, which goes round the Sun every 33 years. It last flew past us in February 1998, but was only visible using binoculars. With Temple-Tuttle the meteoroids are all bunched up in one part of the orbit directly behind the comet, which is why good displays occur only about every 33 years. The Leonids' return in November 1999 was watched with great interest, and indeed a storm reaching 5,000 meteors per hour was recorded. Not so, unfortunately, from most of Britain, which was covered in a large blanket of cloud – nothing unusual there. Following years saw greater numbers of Leonids, but nothing approaching the 1966 level, or indeed the views shown in the etchings of earlier spectacles.

Top tips for meteor-shower observing

The Moon can play an annoying part as far as the seeing of individual meteors is concerned. The brighter the Moon, the more the sky is washed with reflected light, hence the more 'shooting stars' become smothered and invisible. So before you go to any great effort organising a meteor watch, find out what the Moon is up to.

Good luck: from now on you should be making at least five wishes an hour. If you're very lucky you could be making over a hundred!

The tremendous Leonid storm of 1833 prompted a frightened observer to exclaim: 'The world is on fire!'

Step-by-step guide to meteor-watching

1

Be prepared. Make it as easy for yourself as possible. Go out (on the right day!) with warm clothing. If you're watching from the garden, set up a deck chair during the day when it's light in a place with the best horizon and away from street lights. You can then go straight there and sit down. Deck chairs are also really good for supporting your neck. As for all astronomy, it's best to use a torch with a red filter (see page 13), as this does not affect your dark-adapted eyes.

2

Make up a flask of something warming. Not too warming, or else the number of meteors you see will dramatically increase beyond plausibility, if you know what I mean. Add a couple of friends and you're in for a good night.

3

Now, if you want to do a meteor-shower watch that will actually be of use to someone, then all you need to do is make a report using all the information detailed in the example below. Make sure you have plenty of pens or pencils when you go out, as they have a tendency to fall on the ground in the dark and then quickly vanish down very small black holes, never to be seen again. Also make sure you have an accurate clock - only correct timings for meteors are useful. Of course, because you have already studied this book in depth, the actual process of locating each meteor within any constellation will be a piece of cake.

An example of an extremely useful meteor report.

A Super Meteor Shower Report

Date: use double-dating (i.e. Aug 12-13 for the night of the 12th)	**Observer(s):** this would be your name then
Observing Site: give latitude and longitude	**Address:** no guesses what goes in here
Watch Times: at least one hour is best	Plenty of room in this bit to do some rambling... Here are two terms from the table below explained:
Sky Conditions: misty, high cloud, moonlight, sparkly, etc.	**Sporadic** means the meteor seen did not belong to the shower you are observing
Star Limiting Magnitude: this is the faintest star you can see	**Trains** are any remaining "vapour trail" remaining after the initial meteor streak

Meteor Number	Time (UT) (same as GMT)	Magnitude	Shower or Sporadic	Constellation(s) (in which the meteor is seen)	Trains	Notes
1	23.45	2	Perseid	Camelopardalis	2 seconds	Yipee - number one!
2	23.46	-1	Perseid	Cassiopeia to Cepheus	-	Crazy Reddish
3	23.51	0	Sporadic	Ursa Minor to Draco	1 second	Where did that come from?
4	23.53	-3	Perseid	Pegasus	2 seconds	Fragmented - A korker!
5	23.58	0	Perseid	Equuleus to Sagittarius	-	Zippo!
6	0.04	-5	Perseid	Andromeda to Lacerta	4 seconds!	Do-dar do-dar day!!!!
7	0.08	-2	Perseid	Pegasus	1 second	Hercules! Boogie, Boogie.

List of yearly meteor showers

Those highlighted in bold are the major showers of the year.

SHOWER	PEAK DATE	DATE RANGE	MAX. NUMBER PER HOUR	NOTES
Quadrantids	**Jan 3**	**Jan 1 – Jan 6**	75	**Yellow and blue meteors. Medium speed.**
α Centaurids	Feb 8	Jan 28 – Feb 21	5 – 20	A southern shower with some very bright, swift meteors.
Aurigids	Feb 5-10	Jan 31 – Feb 23	10	Not one of the best showers.
Virginids	Apr 7-15	Mar 10 – Apr 21	5	Slow, long trails with multiple radiants.
Lyrids	Apr 22	Apr 17 – Apr 25	15	Fastish meteors from Comet Thather 1861 I.
π Aquarids	**May 5**	**Apr 22 – May 20**	50	**A southern shower with very fast, bright meteors. Associated with Halley's comet.**
June Bootids	Jun 27	Jun 27 – Jul 5	Variable	A slow-speed variable shower with occasional strong outbursts. From Comet Pons-Winnecke.
Capricornids	Jul 5-20	Jun 10 – Jul 30	5	Slow, yellow and blue bright meteors. Several peak dates and radiants.
δ Aquarids	Jul 29 Aug 8	Jul 15 – Aug 19	20 5	A southern shower with a double peak.
Piscis Australids	Jul 30	Jul 16 – Aug 12	5	Southern hemisphere shower of slowish meteors.
α Capricornids	Aug 1	Jul 15 – Sep 10	5	Produces slow fireballs visible for many seconds.
ι Aquarids	Aug 6 Aug 25	Jul 15 – Sep 18 Jul 1 – Sep 18	8	Fainter slow-to-medium-speed meteors from a double radiant.
Perseids	**Aug 12**	**Jul 23 – Aug 21**	75	**Many bright meteors with trails from Comet Swift-Tuttle.**
Piscids	Sep 8 Sep 21	Aug 25 – Oct 20	10 5	Several peaks of slower meteors from different radiants.
Giacobinids or Draconids	Oct 8	Oct 6 – Oct 10	Variable	Slow meteors from Comet Giacobini-Zinner.
Orionids	**Oct 21**	**Oct 15 – Oct 29**	25	**Fast with many trails. From Halley's comet.**
Taurids	Nov 4	Oct 15 – Nov 30	10	Bright and slow meteors from Comet Encke.
Leonids	Nov 17	Nov 14 – Nov 20	Variable	Very fast meteors with trails from Comet Temple-Tuttle.
Geminids	**Dec 14**	**Dec 6 – Dec 18**	90	**Medium speed, bright meteors, from asteroid Phaethon (3200).**
Ursids	Dec 22	Dec 17 – Dec 25	10	Slow meteors from Comet Temple-Tuttle.

Let me finish by saying that meteor showers have generally nothing to do with meteorites. Shooting stars are the dusty bits from comets, whereas meteorites are stony, metal or mixtures of both and are mainly chips off the asteroid belt.

A Final View of Everything

And so this book, a tiny part of the Universe if you like, draws inexorably to a close.

'What has this book got to do with the Universe?' I hear you cry.

Well, the Universe is absolutely everything – it includes us living here on our tiny Earth. We are of course, only one of the planets orbiting our Sun, which is actually a star, a glowing ball of gas, similar to many of the stars we can see in the night sky. These and more distant stars, with all the dust and gas in space, form huge groups known as galaxies – and there are literally billions and billions of them. If you were to travel far enough, all these galaxies would seem to live in huge families (known as super-clusters) that weave across space and time in vast sheets and bubbles.

Then we have amazing deep-sky objects: out there are stars that have just used up their fuel and at this very moment are collapsing in on themselves. The rate of collapse is such that the matter gets compressed and gravity becomes so powerful that not even light can escape – a black hole has been made. If we were to imagine the 'shape' of gravity and space as being like a rolling landscape with stars in the dips, then a black hole would be like a vast hole in the ground that drops forever. Not something you'd want to encounter on a walk through the countryside or on a spaceship travelling through the Cosmos. But black holes are just another object found in the Universe – just as this book is. It's a varied place wherever you look.

Eventually life, such as you and me, came about. Eventually *Simple Stargazing* came about – it's all linked, you see.

You probably started with a wish to understand the night sky, and I hope that this book has given you a taste for stargazing and shown that there is so much more to it all. Our ancestors revered their gods, weaving sky-lore into our present calendars. The desire to understand the sky drove the beginnings of science. Theories and our knowledge of the Cosmos have changed ever since as we constantly devise new ways of studying and understanding space and our place in it. As we uncover more secrets, we add more pieces to our Universal jigsaw puzzle, but at the same time the puzzle itself keeps growing. The journey is far from over – it all comes from our quest for knowledge. I hope I've given you plenty of Universal ideas with which to continue your own stargazing voyage of discovery.

Starry skies,

ANTON

AstroGlossary

Asterism
The name we give to an easy-to-remember group of stars generally, but not exclusively, within a constellation. Examples include: the Plough (Ursa Major), the Teapot (Sagittarius) and the magnificent GSSC (see page 82).

Black hole
One type is the remains of a massive old dead star. Fuel, in the form of gas, keeps a stellar furnace shining brightly, but once that fuel runs out the pressure holding the star up stops and the force of gravity takes over – it collapses in on itself. This enormous 'sucking' force of gravity becomes so strong that not even light can escape and it appears black, hence the name. Avoid these if at all possible.

Circumpolar
An object, such as a star or a constellation, that is so close to one of the poles of the sky that it never goes below the horizon. Just as you can circumnavigate the world, meaning that you can sail all the way round it, circumpolar simply means 'going around the pole'.

Ecliptic
A Greek word meaning 'place of eclipses'. This imaginary line is the path that the Sun takes around the sky over the year. The ecliptic passes through 12 constellations, giving rise to our 'signs of the zodiac'. The word *zodiac* itself means 'line of animals', and our word *zoo* comes from the same source – isn't this all such a great font of knowledge! Because the solar system was all made from a disc of space stuff, we find that the Moon and planets also stay close to this line.

Elliptical orbit
An orbit whose path is an oval shape instead of circular. The more elliptical the orbit the more squashed the oval. The Earth has a slightly elliptical orbit around the Sun which means that in January we are 5 million km closer to the Sun than we are in July.

Galaxy
Around one million or more stars, sometimes with cupploads of dust and heaps of gas, all held together by gravity in many different shapes including spiral and elliptical. When we use the word Galaxy with a capital G we mean our home Galaxy – the one that our Sun, and therefore we, are a part of.

Libration
An effect that allows us to see 59 per cent of the Moon's surface in total – though only 50 per cent at any one time. It is the result of several factors, including the Moon's orbit being tilted and slightly elliptical around the Earth as well as the change in our viewing position throughout the day.

Light-year
The distance that light travels in one year. This is 9,460,530,000,000 (9.5 trillion) kilometres.

Magnitude
An object's apparent brightness in the sky is called its apparent or visual magnitude.

Occultation
The time when one object blocks the light from another, such as the Moon moving in front of Jupiter. If a star or planet catches the edge of the Moon, this is known as a grazing lunar occultation.

Parallax

A way of finding out how far away things are using positions of the Earth's orbit around the Sun at different times to detect tiny movements in nearby stars. Just like holding your finger up at arm's length, then seeing how the background changes as you look at your finger using your left and then your right eye. You see a shift in the position of your finger against the background, whereas astronomers detect the shift in the position of a nearby star against the more distant background ones. Then add maths.

Planet

A ball of rock, gas or ice that orbits a star. The planets in our solar system do not make their own light, as stars do; we see them shining away because they reflect the light from the Sun. Generally, an object should be in the order of 2,000 km or more in diameter to be classed as a planet, but the debate rages as to the exact size.

Pole star

The name given to any bright star that happens to be more or less directly above the Earth's north or south pole. At present Polaris is the northern Pole Star, while nothing of note marks the southern pole. Because the Earth wobbles on its axis over 25,800 years, the axis slowly points to different parts of the sky and so different stars can become the Pole Star. In the northern hemisphere we find that (among many others) Vega (in Lyra) and Thuban (in Draco) will be (and already have been) the Pole Star, while the southern equivalents include Alpha Hydri (in Hydrus) and Gamma Velorum (in Vela).

Proper motion

Now, just when you thought it was safe to learn the night sky, there's an added difficulty of moving stars. Indeed, these 'fixed' stars of the Greeks are anything but - I wonder how they wander. All stars are racing through the heavens; it's just that some are moving quite differently to the Sun, and they give their motions away by slowly changing their positions over time. This extremely tiny movement, known as the *proper motion*, is given in arc seconds per year.

Radiant

The point in the sky where meteors seem to be coming, or radiating, from - quite clever, you see. Of course, meteors can appear anywhere but if you trace each trail back it will be the radiant you find - presuming that all the meteors you see are a part of that shower.

Sidereal month

The time it takes for the Moon to orbit the Earth relative to the starry background: 27.3 days. If the Earth was not orbiting the Sun, then Moon phases would repeat after 27.3 days, but they don't, and this is what gives rise to the **synodic month**.

Solar system

Our 'solar system' includes the Sun and everything that is held by its gravity - the nine known planets, their moons, the asteroids and comets. There are other solar systems out there, made up of planets that we have found orbiting other stars. There may be other worlds, but at present our solar system is the only place where we know life exists.

Star

A great ball of hot gas that shines because of nuclear reactions deep down inside it. You need 10 million degrees Celsius to get a star going - then stand back!

Supernova

A star that ends its days with a massive gigantenourmongous explosion. A supernova is a variable star to end all variable stars – a one-use-only variety. The tremendous brightening is caused by a final gravitational collapse and sensational rebound. If you've got one, be careful!

Synodic month

The time it takes for the Moon to orbit the Earth relative to the Sun. This is 29.5 days and is also the time it takes for Moon phases to repeat themselves; so two successive Full Moons are 29.5 days apart.

Universal time (UT)

Sounds very impressive. Scientists have been fooling us for years with astro-time-speak like this. It is simply the space word for GMT, or Greenwich Mean Time. More complicatedly, there are a number of varieties of UT including UT0, UT1 and UTC – those scientists and their whacky names!

Waning Moon

The phases of the Moon from Full to New. The waning Moon first becomes visible in its gibbous phase (between half and full) just after the Sun has set, but on the opposite side of the sky. Day by day the Moon moves towards the morning Sun, shrinking past half to crescent, and finally disappears as the New Moon.

Waxing Moon

The phases of the Moon from New to Full, and it's this waxing stage that most of us see. The Moon first becomes visible as a thin crescent in the evening sky close to the setting Sun. As the days go by, the Moon moves away from the Sun and grows to a half, gibbous, then Full Moon.

Going Further

Further Reading

Dunlop, Storm and Wil Tirion, *Collins Atlas of the Night Sky* (Collins, 2005)

Dunlop, Storm and Wil Tirion, *How to Identify the Night Sky* (Collins, 2004)

Dunlop, Storm and Wil Tirion, *Wild Guide Night Sky* (Collins, 2004)

Illingworth, Valerie, *Collins Dictionary of Astronomy*, 2nd edn (Collins, 2000)

Moore, Patrick, *The Guinness Book of Astronomy* (Guinness, 1988)

Moore, Patrick, *Philip's Atlas of the Universe* (Philip's, 2003)

Ridpath, Ian and Wil Tirion, *Collins Pocket Guide Stars and Planets*, 3rd edn (Collins, 2001)

Ridpath, Ian, *Norton's Star Atlas and Reference Guide* (Longman, 1998)

Tirion, Wil, *Cambridge Star Atlas* (Cambridge University Press, 2001)

I found the following old books and resources extremely useful. They are all out of print, but you might have fun trying to track them down.

Allen, Richard Hinckley, *Star Names, Their Lore and Meaning* (Dover Publications, 1899)

A Lady (genuinely!), *Urania's Mirror* (Samuel Leigh, c.1825)

Astronomical Images on Card (J. Reynolds & Son, c. 1860)

Heath, Thomas, *Twentieth Century Atlas of Popular Astronomy* (W. & A. K. Johnston Ltd, 1903)

Mee, Arthur, *Observational Astronomy* (Western Mail Ltd, 1893)

Milner, Rev. Thomas, *The Gallery of Nature* (W. & R. Chambers, 1846)

Magazines (available in the UK)

Astronomy, Astro Media Corp., Milwaukee

The Sky at Night, Bristol Magazines, Bristol

Sky & Telescope, Sky Publishing Corp., Cambridge, Mass.

Societies

British Astronomical Association, Burlington House, Piccadilly, London, W1J 0DU
http://www.ast.cam.ac.uk/~baa

Society for Popular Astronomy, 36 Fairway, Kayworth, Nottingham, NG12 5DU
http://www.popastro.com/home.htm

Royal Astronomical Society of Canada, 136 Dupont Street, Toronto, Ontario, M5R 1V2
http://www.rasc.ca

Royal Astronomical Society, Burlington House, Piccadilly, London, W1J 0DU
http://www.ras.org.uk

Multimedia (CD Rom)

Maris Multimedia, *Redshift 4* (Dorling Kindersley, 2000)

Websites

THE MOON
NASA images, missions and facts:

http://www.nssdc.gsfc.nasa.gov/planetary/planets/moonpage.html

The Hitchhiker's Guide to the Moon for maps of the entire surface:

http://www.shallowsky.com/moon/hitchhiker.html

ECLIPSES
NASA's definitive list of forthcoming eclipses:

http://www.sunearth.gsfc.nasa.gov/eclipse/eclipse.html

THE PLANETS
An in-depth look at all the solar system has to offer:

http://www.nineplanets.org

STARS
All you ever wanted to know about stars, including star of the week!

http://www.astro.uiuc.edu/~kaler/sow/sowlist.html

THE MILKY WAY
An interactive journey that enables you to see and understand our Galaxy:

http://www.anzwers.org/free/universe

SATELLITES AND THE ISS
Huge numbers of satellites are visible from your back garden. Here's where to go to find out when to look:

http://www.heavens-above.com

COMETS
When any comets become visible, here is where you'll find the news:

http://www.encke.jpl.nasa.gov

METEORS
More detailed information on all the showers of the year:

http://www.comets.amsmeteors.org/meteors/calendar.html

VESTA
When Vesta becomes visible the details will be here:

http://www.skyandtelescope.com/observing/ataglance

LATEST NEWS
All the news about what's happening in the sky now as well as plenty of other starry stuff from my site:

http://www.vamplew.co.uk

Places to visit in the UK

There are plenty of places to visit around the country to increase your space knowledge. Here is just a selection:

Herstmonceux Science Centre is a discovery centre in East Sussex where you can experience science through hands-on activities. The centre is set in the grounds of the Old Royal Observatory after it moved from Greenwich, with the fine large telescopes open for general viewing.
LOCATION: Herstmonceux, East Sussex
http://www.the-observatory.org

Otford Solar System Walk gives you an idea about the scale of the solar system. On the recreation ground you find the Sun, Mercury, Venus, Earth, Mars and Jupiter. Then it's an expedition around the village looking for Saturn, Uranus, Neptune and Pluto. All the planets are at their correct relative sizes, distances and positions to each other (more or less!) as on 1 January 2000 – it's a millennium solar system walk. Great for all trying to get a grip on how big the solar system really is.
LOCATION: Otford, Sevenoaks, Kent
http://www.otford.org/solarsystem

Royal Observatory, Greenwich was founded in 1675 by King Charles II, and built specifically as an aid to navigation. This is where the Astronomers Royal lived and worked until the 1950s, when they were forced to move to Herstmonceux because of light pollution. The Royal Observatory is home to the Prime Meridian of the World, Longitude 0 – this is the centre for all time and space. You'll find the largest optical telescope in Great Britain and live planetarium shows within the historic surroundings.
LOCATION: Greenwich Park, London
http://www.rog.nmm.ac.uk

Bristol Science Centre allows you to step inside a mega-shiny steel sphere and be transported into the future. Recline in one of 100 seats and be immersed in a 3D virtual world. Here you can explore the

Universe by travelling to the far-reaches of space, and learn about the night sky.

LOCATION: Bristol

http://www.at-bristol.org.uk

National Space Science Centre will take you on a journey of discovery through interactive challenges, images, sounds and real space hardware. The stories, people and technology of the past and present will combine to explain our current understanding of space and how it will affect our future. The centre also houses a planetarium.

LOCATION: Leicester

http://www.nssc.co.uk

Techniquest is a science discovery centre with 160 hands-on exhibits: puzzles, challenges and scientific marvels to enthuse and amuse. Then there's a planetarium, a lab, a discovery room and a hi-tech science theatre.

LOCATION: Cardiff Bay

http://www.techniquest.org

Armagh Planetarium offers star shows and talks on astronomy to schools in Northern Ireland. Regular telescope nights are held during the year when members of the public can visit the planetarium and use Ireland's largest public telescope to view some of the breathtaking sights in the night sky.

LOCATION: Armagh

http://www.armaghplanet.com

Glasgow Science Centre is a modern, hands-on, 'learn about the world' kind of place. It houses a marvellous planetarium with one of the finest night skies you'll ever see.

LOCATION: Glasgow

There are also many portable planetaria that visit schools. **Astrodome** is one such company, specialising in teaching the relevant space curriculum areas to infant, junior and senior schools alike whilst students sit amongst the stars.

http://www.astrodome.tv

Details of your local Astronomical Society can be obtained from your local library's reference section.

Index

Page numbers in *italic* refer to the illustrations

Acknowledgements

Special thanks to my wife Gillian, Greg Smye-Rumsby and Peter Michaud. My star planetary images were taken using an Epson 3000z PC digital camera. Thanks to Janice Gibson at Epson UK for the use of this camera.

Collins